Chris Defonseka

Water-Blown Cellular Polymers

Also of interest

Shape Memory Polymers
Theory and Application
Kalita, 2018
ISBN 978-3-11-056932-2, e-ISBN 978-3-11-057017-5

Polymeric Composites with Rice Hulls
An Introduction
Defonseka, 2019
ISBN 978-3-11-063968-1, e-ISBN 978-3-11-064320-6

Flexible Polyurethane Foams
A Practical Guide
Defonseka, 2019
ISBN 978-3-11-063958-2, e-ISBN 978-3-11-064318-3

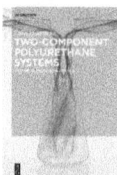

Two-Component Polyurethane Systems
Innovative Processing Methods
Defonseka, 2019
ISBN 978-3-11-063957-5, e-ISBN 978-3-11-064316-9

e-Polymers.
Editor-in-Chief: Seema Agarwal
ISSN 2197-4586
e-ISSN 1618-7229

Chris Defonseka

Water-Blown Cellular Polymers

A Practical Guide

2nd Edition

DE GRUYTER

Author
Chris Defonseka
Toronto
Canada
defonsekachris@rogers.com

ISBN 978-3-11-063950-6
e-ISBN (PDF) 978-3-11-064312-1
e-ISBN (EPUB) 978-3-11-063978-0

Library of Congress Control Number: 2018966603

Bibliographic information published by the Deutsche Nationalbibliothek
The Deutsche Nationalbibliothek lists this publication in the Deutsche Nationalbibliografie;
detailed bibliographic data are available on the Internet at http://dnb.dnb.de.

© 2019 Walter de Gruyter GmbH, Berlin/Boston
Typesetting: Integra Software Services Pvt. Ltd.
Printing and binding: CPI books GmbH, Leck
Cover image: Science Photo Library / EYE OF SCIENCE

www.degruyter.com

Preface

Cellular polymers are an important branch of plastics and have an essential role in daily life. They are also known as 'foamed polymers', 'polymeric foams' or 'expanded plastics'. In a world changing constantly to reduce costs, researchers and scientists have been looking for lighter (yet strong) materials, especially for automobile, aircraft, building-construction and packaging industries. With regard to plastics applications, the shift from solid polymers to cellular polymers has been the answer.

Most plastic polymers can be foamed but only relatively few of them have commercial importance: polystyrenes (PS), polyolefin, polyvinyl chloride, polyimides and polyurethanes (PU). Of these, two stand out clearly: PS for packaging and insulation and PU for comfort. Blowing agents have a major role in the manufacture of expanded or cellular polymers. Since the birth of the concept of the need for cellular polymers, the blowing agents used have been petroleum-based, which results in harmful gases being emitted into the atmosphere. For several years, environmental concerns have resulted in laws to prevent some of these harmful blowing agents being used.

Blowing agents having lesser impacts on environmental issues have been allowed, but researchers have created a novel 'blowing' or 'expansion' system for PS and PU – water-blown cellular polymers. This is a very exciting concept and very encouraging for overcoming environmental concerns because three of the largest industrial applications are based on these cellular polymers. Previously, this concept was confined to only petroleum-based polymers but, due to diligent and constant research, scientists have created water-blown systems, even for polymers based on natural oil-based polymers. Moreover, these blowing systems are also being used for specialty polymers to produce cellular polymers.

The polymerisation stages and processing methods may differ from those using conventional approaches. Moulding of these water-blown cellular polymers may also have some issues, with the consequent advantages and disadvantages. This book discusses these issues in detail and also offers recommendations for machinery, equipment and processing parameters to obtain optimal results.

The readers of this book will benefit from the in-depth knowledge and information provided by the author, who is active as a consultant in local and international arenas. He imparts detailed knowledge from theoretical and very valuable practical experience from having rendered assistance to producers and moulders (small and large) in several countries. This book provides a thorough understanding of water-blown cellular polymers and how to process and mould them. As such, this book is ideally suited for students, entrepreneurs, teachers, professionals as well as for small and large moulders of water-blown cellular polymers.

https://doi.org/10.1515/9783110643121-201

Contents

1 What are polymers?

A polymer is a large molecule ('macromolecule') composed of structural units typically connected by covalent chemical bonds. The basic unit of a polymer is a *mer* and *poly* means 'many', so the word 'polymer' means a substance with many basic units. There are different types of mers and, when they are joined by a process called polymerisation, three basic types of polymer are formed.

If the same type of mers is joined together, they are called 'homopolymers'. If two different types of mers are joined together, they are called 'copolymers'. If three different types of mers are joined together, they will form 'terpolymers'.

These polymers are classified into two main categories: 'thermoforming' (they soften upon heating and after moulding can be re-used) and 'thermosetting' (they cannot be re-used after moulding). Some simple examples of thermoforming polymers are polyethylenes (PE), polystyrenes (PS) and polypropylenes (PP). Examples of thermosetting polymers are silicones, polyurethanes (PU) and melamines.

Because of the extraordinary range of properties of polymeric materials, they have established themselves as an essential part of everyday life, ranging from the very familiar plastics and elastomers to natural biopolymers such as deoxyribonucleic acid and proteins (which are essential for life). Most plastics are based on ethylene gas derived from crude oil. A simple example of a basic plastic is PE, the repeating units of which are based on ethylene monomer. Polymeric materials from natural sources such as shellac, amber and natural rubber have been known for centuries. Biopolymers such as proteins and nucleic acids have crucial roles in biological processes. Other varieties of natural polymers are available (e.g., cellulose is the main constituent of wood and paper).

1.1 Microstructure

The microstructure of a polymer (sometimes called 'configuration') relates to the physical arrangement of monomer residues along the backbone of the polymer chain. These are the elements of polymer structure that require the breaking of the covalent bond to change. A structure has a strong influence on the properties of a polymer.

An important microstructural feature determining polymer properties is the 'architecture' of the polymer. The simplest polymer architecture is a linear chain: a single backbone with no branches. A related non-branching architecture is a ring polymer. A branched polymer molecule is composed of a main branch point in a polymer chain with one or more chains or branches. Branching of polymer chains affects the ability of chains to 'slide' past one another by altering intermolecular forces, which in turn affects the bulk physical properties of a polymer. Long-branch chains may increase polymer strength, toughness and the glass transition temperature (T_g)

https://doi.org/10.1515/9783110643121-001

due to an increase in the number of entanglements per chain. The effect of such long-chain branches on the size of the polymer in solution is characterised by the branching index. Conversely, random lengths and short chains may reduce polymer strength due to disruption of organisation and may likewise reduce the crystallinity of the polymer. A good example of this effect is related to the range of physical attributes of PE. High-density polyethylene (HDPE) has a very low degree of branching, is quite stiff and is used in applications such as containers and milk jugs. Conversely, low-density polyethylene (LDPE) has a significant number of long and short branches, is quite flexible and is used in applications such as plastic films.

The architecture of a polymer is often physically determined by the functionality of the monomers (basic mers) from which it is formed. The property of a monomer is defined as the number of reaction sites at which it may form chemical covalent bonds. The basic functionality required for forming even a linear chain is two bonding sites. Higher functionalities yield branched or even crosslinked or network polymer chains.

An effect related to branching is chemical crosslinking by formation of covalent bonds between chains. Crosslinking tends to increase the T_g and increase strength and toughness. This process is used to strengthen rubber compounds in a process known as 'vulcanisation', which is based on crosslinking with sulfur. For example, to a great degree vulcanisation is essential in applications such as car tyres, whereas rubber erasers are not crosslinked to promote flaking when used, thereby protecting the paper. A crosslink may provide branch points from which four or more distinct chains emanate. A polymer molecule with a high degree of crosslinking is referred to as a 'polymer network'.

1.2 Lengths of polymer chains

Physical properties of a polymer are strongly dependent on the size or length of the polymer chain. For example, as chain length is increased, melting points and boiling points increase, as does the impact strength and viscosity (resistance to flow) of the polymer in its melt state. A tenfold increase in the length of a polymer chain results in a viscosity increase of >1,000-fold. Increasing chain length tends to decrease chain mobility, increase strength and toughness, and increase the T_g. A common means of expressing the chain length of a polymer is the 'degree of polymerisation', which quantifies the number of monomers incorporated into the chain. As with other molecules, the size of a polymer may also be expressed in terms of molecular weight (MW).

1.3 Polymer morphology

Polymer morphology can, in general, be described as the arrangement chains and the order of the many polymer chains. A synthetic polymer may be described as 'crystalline' if it contains regions of three-dimensional ordering on scales of atomic (rather than macromolecular) length, usually arising from intramolecular folding and/or stacking of adjacent chains. Synthetic polymers may consist of crystalline and amorphous regions, the degree of crystallinity being expressed in terms of a weight fraction or volume fraction of crystalline material.

The crystallinity of polymers is characterised by their degree of crystallinity, ranging from 'zero' for a completely non-crystalline polymer to 'one' for a completely crystalline polymer. Polymers with microcrystalline regions are, in general, tougher, can be bent more without breaking and more impact-resistant than totally amorphous polymers. Polymers with a degree of crystallinity approaching zero or one tend to be transparent, whereas polymers with intermediate degrees of crystallinity tend to be opaque due to light scattering by crystalline or glassy regions. Thus, for many polymers, reduced crystallinity may also be associated with increased transparency.

1.4 Mechanical properties

The tensile strength of a material quantifies how much stress the material can endure before undergoing permanent deformation. Tensile strength is very important in applications that rely on the physical strength or durability of a polymer. For example, a polymer with high tensile strength will hold a greater weight before snapping than a polymer with low tensile strength. In general, tensile strength increases with the length of the polymer chain and crosslinking of polymer chains. Young's modulus quantifies the elasticity of a polymer, and is the ratio of rate of change of stress to strain. Like tensile strength, Young's modulus is very important in polymer applications involving the physical properties of polymers, and is strongly dependent on temperature.

1.5 Phase behaviour

In general, the melting point of a material would suggest solid-to-liquid transition. However, if applied to polymers, the melting point of a material is more a transition from a crystalline or semi-crystalline phase to a solid-amorphous phase. This property is more appropriately called the 'crystalline melting temperature'. Among synthetic polymers, crystalline melting is discussed only with respect to thermoplastics because thermosetting polymers decompose at high temperatures rather than melt. The boiling point of a polymeric material is strongly dependent on chain length. Polymers with a large degree of polymerisation do not exhibit a boiling point because they decompose before reaching theoretical boiling temperatures. For shorter chain lengths, a boiling transition may be observed.

A parameter of particular interest in manufacture of synthetic polymers is the T_g, which describes the temperature at which amorphous polymers undergo a second-order transition from a rubbery, viscous amorphous solid or from a crystalline solid to a brittle, glassy amorphous solid. The T_g may be engineered by altering the degree of branching or crosslinking in the polymer or by the addition of plasticisers.

1.6 Mixing behaviour

In general, polymeric mixtures are far less miscible than materials with small molecules. Usually, the driving force for mixing is entropy, not interaction energy. That is, miscible materials usually form a solution not because their interaction with each other is more favourable than their self-interaction, but because of an increase of entropy, and hence free energy associated with increasing the amount of volume available to each component. Polymeric molecules are much larger and have much larger specific volumes than smaller molecules, so the number of molecules involved in a polymeric mixture is far smaller than the number in a small molecule mixture of equal volume. Furthermore, the phase behaviour of polymer solutions and mixtures is more complex than that of small-molecule mixtures. In dilute solution, the properties of the polymer are characterised by the interaction between the solvent and polymer. In good solvents, a polymer will appear swollen and will occupy a large volume.

Inclusion of plasticisers tends to lower the T_g and increase polymer flexibility. In general, plasticisers are small molecules that are chemically similar to polymers and create gaps between polymer chains for greater mobility and reduced interchain interaction. A good example of the action of plasticisers is related to polyvinyl chloride (PVC). Unplasticised polyvinyl chloride (uPVC) is used for production of pipes, which have no plasticiser in them, because they need to be strong and heat-resistant. If plasticised PVC is used for pipes, they will lose the plasticiser through evaporation due to heat, and will become brittle and split. Plasticised PVC is used for flexible applications such as coatings, sheeting, and films.

1.7 Chemical properties

The attractive forces between polymer chains play a large part in determining the properties of a polymer. Polymer chains are long, so these inter-chain forces are amplified far beyond the attractions between conventional molecules. Different side groups can lead the polymer towards ionic bonding or hydrogen bonding between its own chains. Typically, these stronger forces result in higher tensile strength and higher crystalline melting points.

The intermolecular forces in polymers can be affected by molecules with different charges ('dipoles') in the monomer units. Polymers containing covalent double-bond ('carbonyl') groups can form hydrogen bonds between adjacent chains. These strong hydrogen bonds, for example, result in the high tensile strength and melting points of polymers containing urethane or urea linkages. Dipole bonding is not as strong as hydrogen bonding but produce greater flexibility.

Ethylene, however, has no permanent dipole. The attractive forces between PE chains arise from weak van der Waals forces. Molecules can be thought of as being surrounded by negative electrons. As two polymer chains approach, their electron clouds repel one another. This phenomenon lowers the electron density on one side of a polymer chain, creating a slightly positive dipole on the side, which is sufficient to attract a second polymer chain. The attractive forces are quite weak, so PE can have a lower melting temperature compared with other polymers.

1.8 Polymer degradation

Polymer degradation refers to a change in properties such as colour, tensile strength, shape, and MW. Polymer degradation can be caused by heat, light, chemicals, extreme cold and, in some cases, galvanic action. Often, polymer degradation is due to the hydrolysis (separation due to the action of water) of the bonds connecting the polymer chains, which in turn leads to a decrease in the molecular mass of the polymer. During processing, degradation of polymer materials can take place due to excessive heat or dwell times. Carbon-based polymers are more prone to thermal degradation than inorganically bound polymers and, therefore, are not suitable for most high-temperature applications. However, the degradation process can be useful for understanding polymer structure in terms of recycling/re-use of polymer waste to prevent/reduce environmental pollution.

1.9 Common thermoforming and thermosetting polymers

The full range of these polymers is too vast to mention here but a few of the most common ones are listed in Table 1.1.

These polymer resins can be in the form of powders, granules, beads, pre-pegs and, in some cases, in liquid form. Manufacturers of plastic resins pack them in paper bags (25 kg), drums (200 kg) or for bulk packing in totes (≈400–500 kg). Liquid resins may be available in smaller packs. Three important factors to consider are shelf life, fire hazards and appropriate handling for safety.

Table 1.1: Common thermoforming and thermosetting polymers.

Thermoplastic polymers	Thermosetting polymers
– PE	– PU
– PP	– Melamine
– PVC	formaldehyde
– PS	– Silicones
– ABS	– Epoxy
– TPE	– Phenolic
– Polyester	formaldehyde
– Polyamide	
– PC	

ABS: Acrylonitrile–butadiene–styrene
PC: Polycarbonate
TPE: Thermoplastic elastomers

1.10 Composite polymers

Composites are materials made from two or more constituent materials, with significantly different physical and chemical properties that, if combined, produce a final material with characteristics different from the individual components, while remaining separate and distinct within the structure. These composite materials can be formulated to give superior properties and structural strengths. These properties can also be optimised by addition of suitable additives.

Due to unsteady and escalating resin prices, researchers and developers have created composite polymers. The common resins used are LDPE, HDPE, PP and PVC, and other polymers are being tested. Initially, biomass wastes such as wood flour and sisal have been used. Of these, the most common ones until recently were wood–polymer composites (WPC). However, new technology is using rice hulls as the reinforcing constituent, which results in composite resins with even better properties due to the high content of silica in the rice hulls. The ratio of the resin matrix to the reinforcing constituents can be 60:40 but higher reinforcing constituents are being achieved with advancing technology. This emerging 'family' of composites with such great possibilities can be categorised as 'thermoplastic bio-composites' and is a great scientific achievement. Polymeric composites with rice hulls can produce superb lumber, which are an ideal substitute for natural wood. General polymeric composites with rice hulls (PCRH) products tend to be solid and heavy. If lighter weights or cellular cores are desired, small quantities of blowing agents can be incorporated. Dyes and pigments of master batches can be used for colouring, and special additives give wood-like veneers and matt or glossy finishes. These versatile polymer resins can be processed on conventional and standard machinery but with slightly different processing parameters.

1.11 Some important and common polymers

Here, I present, briefly, five of the most important and commonly used polymers in industry: PE, PS, PP, PVC and PU. The first four are thermoplastic resins, whereas the last is a thermoset. Pentane gas is used as the blowing agent for producing cellular polystyrene [expandable polystyrene (EPS)], but newer grades are using water as the blowing source. This grade is called 'water-blown expandable polystyrene' (WEPS). PU are water-blown polymers and give excellent foams.

1.11.1 Polyethylenes

PE is probably the most popular polymer in the world. It is used to make grocery bags, pipes, films, bottles, children's toys and even bullet-proof vests. For such a versatile polymer, it has a very simple structure, the simplest of all commercial polymers. In general, it is a byproduct resulting from refining crude oil and derives from the resulting ethylene gas.

A molecule of PE is a long-chain of carbon atoms, with two hydrogen atoms attached to each carbon atom. Figure 1.1 shows the structure of PE and the chain is many thousands of atoms long:

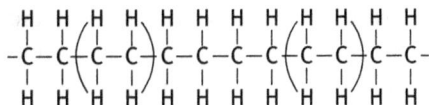

Figure 1.1: PE.

Sometimes, some of the carbons, instead of having hydrogens attached to them, have long-chains of PE attached to them: this arrangement results in LDPE. When there is no branching, linear PE or HDPE will result. Linear PE is much stronger than branched PE, but the latter is easier to produce and less expensive. Usually, linear PE is produced at very high MW (200,000–500,000) and identified as ultra-high-MW PE. The latter is used for making fibres, some of which are so strong and used for bullet-proof vests. PE resins are, in general, available as granules, powders or liquids.

1.11.2 Polystyrenes

PS, a versatile, hard, stiff and brilliantly transparent synthetic polymer resin, is produced by polymerisation of styrene monomer. Styrene is obtained by reacting ethylene with benzene in the presence of aluminium chloride to yield methylbenzene. Then, the benzene group in this compound is dehydrogenated to yield phenyl ethylene or styrene, a clear, liquid hydrocarbon. Then, styrene is polymerised using free-radical initiators primarily in bulk and suspension processes (though solution and emulsion methods can also be employed).

The presence of pendant phenyl groups is key to the properties of PS. Solid PS is transparent owing to these large ring-shaped molecular groups, which prevent polymer chains from packing into close crystalline arrangements. In addition, phenyl rings restrict the rotation of the chains around carbon–carbon bonds, making the polymer rigid and hard. PS can also be copolymerised or blended with other polymers to impart high-impact strength or blended with rubber for more flexibility.

Cellular polystyrene (EPS) is a rigid, closed-cell foamable material. EPS was first made with chlorofluorocarbons (CFC) as blowing agents but the latter have been banned because of environmental concerns. Nowadays, pentane gas is incorporated into PS beads at the time of polymerisation. More recent developments have made it possible to use water as the blowing agent, and these grades are identified as 'WEPS'. These grades produce equally good cellular PS foams as EPS, and contribute considerably to emission of harmless water vapour instead of harmful gases. The only difference in WEPS processing is at the pre-expansion stage of the beads, where slightly different pre-expanders are required.

Extruded polystyrene (XPS) cellular boards are made by extrusion of standard-grade PS resins incorporated with a gas during extrusion. Special extrusion dies are used with auto-width/thickness calibrators as the hot foam mass exits the die into the atmosphere. These boards are, in general, made in widths of ≈1 m and cut-off to desired lengths on the take-off conveyor. Different colours are made using dyes or masterbatches to identify thermal-conductivity, moisture-resistant, and fire-retardant factors.

Standard grades of PS are used for making many items for the domestic market, EPS, and WEPS for making items for packaging, the fishing industry, and for building construction. Modified grades with synthetic rubber for example ABS are used for the manufacture of pipes, computer housings, refrigerator housing and many other applications.

1.11.3 Polypropylenes

PP is a synthetic resin made by polymerisation of propylene. As an important polymer of the polyolefin family, PP can be moulded, blown into film or extruded into many plastics products. Some of their basic properties are: toughness, flexibility, light weight, heat resistance and resiliency. They are used for making metalised film for packaging and are used as matrices for composite resins (which are now becoming popular among manufacturers of plastics products). PP are very versatile polymers and can be spun into fibres for use in industrial and household textiles. Propylene can also be polymerised with ethylene to produce an elastic ethylene-propylene copolymer.

Propylene is a gaseous compound resulting from thermal cracking of ethane, propane, butane or naphtha originating from petroleum fractions. Like ethylene, it belongs to a 'lower olefin class' of hydrocarbons, whose molecules contain a single pair of carbon atoms linked by a double bond. Under the action of polymerisation catalysts, however, the double bonds can be broken and thousands of propylene molecules linked together to form chain-like polymers. These polymers share some of the properties of PE but are stronger, stiffer and harder, and soften at higher temperatures. They are more prone to oxidation than PE unless appropriate stabilisers and antioxidants are added.

PP are blow-moulded into bottles for foods, shampoos and other household liquids. They are also injection-moulded into many products: including appliance housings, dishwater-safe food containers, toys, automobile battery cases, and outdoor furniture. If a thin section of moulded PP is flexed repeatedly, a molecular structure is formed that can withstand additional flexing without cracking. This fatigue-resistance property has led to designs of boxes with PP and other containers, in which 'self-hinged' covers are required. A large proportion of PP polymers are made into yarn for textiles/ropes and into films for packaging. Other applications can be cordage, non-woven fabrics for protective wear, medical applications, and replacement for natural grass in playing fields as fibres, outdoor agricultural applications, and reinforcement in construction/ road paving. These applications take advantage of the excellent properties of PP such as toughness, resilience, water-resistance and chemical inertness.

Expanded (or cellular) PP are also very popular and have an expanding global market. These foamed materials are used widely in packaging, bedding, furniture, building construction and carpet underlay.

1.11.4 Polyvinyl chloride

PVC is a well-known and widely used polymer in domestic, industrial, building construction and agricultural sectors. A polymer resin made from polymerisation of vinyl chloride is required for products in daily life as much as PE, PS and PU. A lightweight rigid plastic in its pure form, it is also manufactured in a flexible 'plasticised' form. Vinyl chloride (also known as 'chloroethylene') is usually obtained by reacting ethylene with oxygen and hydrogen chloride using a copper catalyst. It is a toxic and carcinogenic gas that is handled under special protective procedures. PVC is obtained by subjecting vinyl chloride to highly reactive compounds known as 'free-radicals initiators'. Under the action of these initiators, the double-bond in the vinyl chloride monomers is opened and one of the resultant single bonds is used to link together thousands of vinyl-chloride monomers to form the repeating units of the polymers, which are large, multiple-unit molecules. uPVC is used for rigid applications and plasticised PVC (soft resins) for flexible applications.

Pure PVC is used in applications in the construction trades, where its rigidity, strength and flame-retardant properties are useful in pipes, conduits, siding windows frames and door frames. Because of its rigidity it must be moulded or extruded at >100 °C to initiate decomposition with emission of hydrogen-chloride gas. Decomposition can be reduced by the addition of stabilisers, which are made of metals such as cadmium, zinc tin or lead. Rigidity can be altered by addition of plasticisers, and highly flexible PVC plastisols can be made (depending on the application).

PVC can be: blow-moulded into transparent bottles; made into sheets/films by extrusion blowing; cast into floating devices; used as insulation coating for wires and electrical cables; used for coatings in artificial leather; made into floor tiles, dip coating as decorative or protective coatings, and extruded profiles for the building construction trades; made into fishing floats. PVC is also one of the very few polymers used as the matrix in polymeric-reinforced composites resins in WPC or PCRH.

1.11.5 Polyurethanes

PU can be divided broadly into two basic categories: flexible and rigid. They are, in general, made by reacting a polyol with an isocyanate. Polyols can be based on polyether or polyester. To produce a urethane, other ingredients such as water (blowing agent), stabiliser (tin), surfactant (silicone), filler (calcium carbonate), catalysts and dye/pigments (colour) are used in a formulation. If very soft foams are desired, an additional blowing agent such as methylene chloride can be used. This is an exothermic (heat giving) reaction and precautions must be taken. Three standard manufacturing methods are used: manual foaming, semi-automatic foaming (single blocks); continuous foaming. The foam mass is laid on a slow-moving conveyor and, after semi-cure, the desired size of blocks are cut off. Regardless of the method used, the foamed blocks need a further minimum 24 h of cure before cutting.

Polyether and polyester give soft, open-cell foams and, with the incorporation of additives, special properties (e.g., colour, ultraviolet protection, high resiliency, fire retardancy, anti-fungal, different densities) can be achieved. Polyether polyols produce soft foams for comfort applications, whereas polyester polyols give less flexible and rigid foams due to their smaller cell structure, thus having tighter air flow. Due to environmental concerns, scientists have created 'eco-polyols', from example, from soya. Foams made from these polyols are equally good, but they tend to yield lesser volumes as compared with petroleum-based polyols.

The continued manufacture and use of polyols from petroleum-based products have raised environmental concerns. Hence, priorities have been to move away from petroleum-based polyols and move towards natural oils and other alternatives. Natural oil polyols (NOP; also known as 'bio-polyols') are derived from vegetable oils using different methods. Foams from NOP have similar sources and applications but the materials themselves can be quite different. Their viscosities are also variable and are usually a function of their MW and average number of hydroxyl groups per molecule. Odours are similar to the parent source and differ from polyol to polyol. These can be clear liquids ranging from colourless to medium yellow.

Some of the natural oil-based sources being used to produce polyols are oils from soya beans, rapeseed (canola), jatropha plants and castor beans. Initially, producers of polyols from these sources used them in combination with petroleum-based polyols, the latter being in larger proportions. Now, due to the rapid development of technology, petroleum-based polyols have been phased out. Polyols from 100% oil sources are now being used widely with good results.

Use of waterborne polyurethanes (WPU) has been carried out for some time, and eased concerns about health and environment. WPU are fully reacted PU in a continuous water-phase, resulting in a solvent-free medium. Technological advances suggest that WPU could be substitutes for solvent-based counterparts. The wide array of properties (e.g., high resistance to abrasion, high impact strength, low temperature flexibility) has increased the range of applications.

WPU are being used in several applications (e.g., coating, sealants, adhesives, elastomers). The manufacturing process necessitates a skilled workforce and the products are relatively expensive. This factor, though a deterrent and restraint to growth of this market initially, has been negated by awareness of the many special properties of WPU, and has created a rapidly growing market. The main players in this market are North America, Germany, China and Japan, and the market is expected to reach US$10.0 billion by 2019.

1.12 Processing methods for polymers

Instead of presenting the standard processing methodologies with which readers are probably familiar with, I present facets of processing based on efficiency, cost savings and waste-reduction areas. There are many processing methods for polymers, but only four of the most popular and practised processes are presented.

1.12.1 Blown film

Blown films are an important and essential commodity for packaging. PE [LDPE, HDPE, linear low-density polyethylene (LLDPE)], PVC, PP are commonly used polymers for films, but others are also used. Two of the specialty films are shrink packaging (for bottles, food) and stretch film (for bulk/pallet wrapping). Some of the important parameters are thickness, width, weight, colour and strength. In the case of multiple layer films (as in food packaging), compatibility of substrates and surface adhesion also becomes important. Metalised films are also an important product in packaging applications: polymer films are surfaced with very thin aluminium coatings by coating, lamination or vacuum spraying to form protective barriers (especially for moisture and heat). For cost estimates for metalised films based on coating of one side, an additional cost of 30–40% on total film costs (easiest being by weight) must be initiated.

The basic features for cost efficiency are: minimisation of start-up and shutdown waste; controlling and maintaining desired uniform thickness throughout; prevention of impurities in the resin melt; minimisation of downtime and width control. If in-house facilities are available for recycling of waste, suitable regrind can be fed back directly into a process and non-suitable regrind used for other products. This strategy reduces material costs. For maximisation of efficiency, minimisation of downtime, and cost reduction of the process, blown film lines should be run for as long as possible (perhaps a few days at a time continuously) but this approach requires good maintenance and 24-h operations.

Computers and software packages do most of the work, but running a blown film in advance, before setting up machine parameters for a profitable operation, is a good approach. This strategy can be achieved by understanding and using the formula given in Table 1.2.

The following information can be used as a guideline for film thickness:
- Lightweight: 1.0–1.5 mm
- Most common: 1.6–2.0 mm
- Heavy duty: 3.0–4.0 mm
- Extra heavy duty: >6.0 mm

1.13 Blown shrink films for bottle packaging

Shrink film are, in general, made from polyolefins (PE), PVC, polyethylene terephthalate (PET), PP and others. General production methods are film blowing, extrusion, and casting. For bottle packaging, films are made by an extrusion/blowing and, depending on the end-packaging needs, the film can be single-layered or multiple-layered, so multiple extruders are connected to the blowing head/ring.

For simple bottle packaging, single heat-shrink films are used widely. On the basis of an extrusion/blowing process, wastage of material will occur due to: pre-start purge; start-up of film until the correct thickness of film is achieved; shutdown time. There is almost zero waste when the machine is in operation, However, waste can occur due to power failures as well as breakages of the film due to foreign material or variations in thickness (perhaps due to power surges). Modern machines have preventive devices

Table 1.2: Machine parameters for film production (recommended basic formulations).

Resins	LDPE/LLDPE/HDPE
Film width	500–1,000 mm
Film thickness	0.015–0.080 mm
Production output	100 kg/h
Screw ratio	L:D = 28.1
Screw speed	25–150 rpm
Main drive motor	15 kW
Heating capacity	23 kW
Temperature controls	2 × 6 zones
Barrel cooling fans	1/7 kW × 6
Die head diameter	100–180 mm (adjustable)
Air ring blower	7.5 HP (5.6 kW)
Die heater capacity	13 kW
Take-up roller width	1,100 mm
Take-up motor	2 HP (1.5 kW)
Take-up speed	60 m/min
Winding motor	2 HP (1.5 kW)

These parameters should assist as a guideline in (a) selection of suitable extrusion systems for a particular range of producing films and (b) give an idea of machinery processing costs of operations. One would add other costs (e.g., material, labour, administration, marketing, taxes, and margins) to arrive at total costs.

built-in to counter such occurrences. To minimise waste during start-up and shutdown, production runs are undertaken continuously for as long as possible. Estimated wastage factor would be approximately 4–6% but this could be less where manufacturers have in-house recycling facilities to enable re-use of waste material.

For heat-shrink film packaging for bottles, the data shown in Table 1.3 are useful as a basis for selection of the correct grade of film for a particular application.

Table 1.3: Heat-shrink films – general specifications (typical).

Width	100–2,400 mm
Thickness	20–250 μm
Shrinkage	Transverse (width) >5–50%
	Machine length >10–80%
Tensile strength	Transverse (width) >12 MPa
	Machine length >12 MPa
Shrinkage temperature	150–220 °C
Breaking elongation	>200%
Specific weight	0.93 g/cm^3

1.14 Injection moulding

Injection moulding is probably the largest processing area in plastics. Injection moulding is the production of solid plastic parts in which a hot polymer melt is injected under pressure into a two-part mould, with a single or multiple cavities. Common plastic polymers used are: PE, HDPE, Nylon, PP, TPE, PC, and PS. General standard machine types are: horizontal; vertical; small manually operated ones with hydraulics, toggle modules, all-electric, hybrid and robotic operations. Two pre-set systems (open-loop and closed-loop) are used, and machines can operate on manual, semi-automatic or fully automatic modes. Standard injection moulding comprises solid hot melts but gases can be injected into them, especially in the injection production of structural foams. Manufacturers of plastics products, in general, adopt quality control (QC) systems for quality assurance, and two of the common ones are American Society for Testing and Materials (ASTM) and International Organization for Standardization (ISO). One of the most common QC systems used is statistical process control (SPC). Injection-moulding machines are selected primarily based on 'clamp force' and 'shot weight'. If a machine is said to have a shot weight of, for example 50 g, a buyer should calculate it as having a capacity of 70% of it (i.e., ≈35 g). This calculation is because the machine for operation is set on an 'injecting cushion' (distance between nozzle opening and screw tip). This cushion determines the final weight of an injected part.

When customers request for quotes, four basic factors must be considered: material costs, mould costs, machine time, and overheads. Sometimes, a customer may offer a mould or moulds. Experienced moulding businesses have a system of quoting instantly based on the weight of the customer's sample plus a 'safe percentage'. From experience/performance, a moulder would have worked out the cost per kilogramme and will base the quote on the weight of a sample plus an added margin to cover profit (which will also depend on the volumes involved). This value could be used if a moulder thinks a customer would look elsewhere, or if a quote is delayed.

More accurate estimates can be calculated using the cost calculator shown in Table 1.4 as a guideline.

1.15 Processing of polymer foams

Many types of polymers can be made into foams. These can take the form of flexible, semi-rigid and rigid foams. Of these, polyurethane foams (PUF), PS foams and PVC foams stand out as the most popular and having the widest ranges of applications. Because of their chemical nature, water can have a leading role as the blowing agent for producing foams.

Table 1.4: Recommended calculator of part costs.

Part name:	Job No:	Date:	Drawing No:	Quantity required:
Customer:	Machine:		Material:	QC system:

Gross hourly rate	Parts
Net hourly output	Parts minus rejects
Moulding cycles	Cycles per h
Net daily output	Parts
Total moulded	Parts
Net weekly output	Parts
Total material used	kg
Unit material used	kg/parts
Net production per shift	kg/8 h
Net monthly output	Parts
Gross annual output	Parts
Net annual output	Parts
Overhead costs per part	$
Mould cost per part	$
Machine cost per part	$
Material cost per part	$
Masterbatch cost per part	$
Regrind cost per part	$
Net cost per part	$
Net cost per 1,000	$
Add overheads per 1,000	$
Estimated total cost per 1,000	$
Agreed target price per part	$
Net production cost per part	$
Projected profit	%

This is only a very basic formula for cost calculation. Taking the agreed selling price and targeted cost price based on desired profit margin, these projected data could be used for setting operating parameters for a machine. For example, knowing the targeted or allowable unit material weight, the injecting 'cushion' can be set to minimise waste or over-injection of material. Hence, the operating parameters of the machine can also be set to achieve a suitable hot melt and also a desirable moulding cycle. If a SPC system of quality-control and closed-loop system is used, it is possible to achieve set targets because the system has auto-correct functions.

PUF are, in general, produced using polyols and isocyanates. The primary blowing agent is water for any formula used. To obtain satisfactory foams and to achieve the desired end results, one must have a good understanding of each component and its function. Producers can use a two-component system or the more conventional multiple-component systems, whereby you can produce foams with different properties using the same group of components. Efficiency in a foaming operation starts with good storage facilities for components, correct calculations for formulation, efficient processing to minimise waste, and efficient conversion to end products. Components are highly toxic and hazardous, so safety factors for handling (and even for post-cure stages) are important. Production of PU foam has an inherent waste factor, and should be kept ≈10–15% but some operation wastes are a ≤30–35%, which is totally unacceptable. Most foam wastes are shredded and made into other products so a foam producer could expect a 'kickback' to lower costs. Waste foam can be recycled into carpet underlay, mattress bases, adhesives, varnishes and lacquers, wood preservatives, and pillows.

The recommended cost calculation for PU foams is shown in Table 1.5.

Table 1.5: Template for cost calculation – example for density of ≥23 kg/m^3.

Raw material	kg (a)	Cost/kg (b)	Total cost (a × b)
Polyol	100		
Toulene diisocyanate	47		
Methelyene chloride	3.6		
Water	2.98		
Silicone	0.92		
Tin	0.21		
Amine	0.24		
Colour	0.30		
Total weight	155.25	**Total cost**	X
Weight loss (gas) 5%	7.76		
Net weight	147.49		
Less skins 10%	14.75		
Good foam weight	132.74		
Cost/kg			X/132.74

Adapted from C. Defonseka in *Practical Guide to Flexible Polyurethane Foams*, Smithers Rapra, Shawbury, UK, 2013, p.93 [1]

Regardless of the method used for making foam blocks, the top of each foam block will consist of a rounded top skin. Some processes have devices to eliminate this flat top and reduce foam loss by ≥4%.

1.16 Blow moulding (hollow containers)

Hollow containers of various sizes and shapes are required for packaging. Blow moulding is a two-way operation in which a hot melt is first extruded to make a downward thick-walled 'parison', which is then clamped between two halves of a mould and then blown against the cavity by air. Using this method, small-to-very-large containers can be made. Some of the common polymer resins used are PE,

PVC, PP, PC, PS, and PET. This process can be semi-automatic or fully automatic, with single or multiple moulding stations, moving in a circulatory movement.

For small bottles, producers use an injection-blowing method. For production of multiple layers, the final parison before blowing has several thin layers making up the final thickness of the parison. For an efficient operation, the start-up, shutdown and the material lost in purging should be minimised. Foreign matter in polymer resins should be eliminated completely and, during production runs, the temperature parameters set must be maintained without fluctuation to ensure a well-homogenised hot melt. The thicknesses of the parison wall should be continuously uniform and the air-blowing pressure also should be constant for each blow. Mould cavities should last for only a limited number of mouldings and, if used beyond this limit, may result in surface flaws. Moulds are very costly, so maintenance of efficiency is essential.

For bottles, the contents create high pressure. For example, for soda bottles, a moulder would use a method called 'stretch blow moulding', which is similar to injection blow moulding except that the parison is stretched before blowing to give strength.

High-impact polymers or PC can be used to achieve stronger-walled bottles. PET resin is used widely for applications that need transparency, stiffness, and good tensile strength (e.g., water bottles).

1.17 Extrusion

Extrusion is used for production of solid, hollow or shaped profiles. A very big market exists for these products. Foamed expanded PS boards (XPS for insulation of buildings) and the latest polymer/bio-composite lumber (ideal substitute for natural wood) is made this way. Coated wires (e.g., electrical cables) are made by extrusion and have a huge market. Extrusion is a very interesting and versatile process used to make a wide range of products essential for the building construction industry.

Efficiency starts with correct selection of extruder system/systems, and use of correct screws for maximum output of homogenised hot melts and downstream equipment. Cost efficiency primarily requires well-trained operators, minimisation of start-ups and shutdown times, as well as good maintenance to prevent machine breakdowns. Downtime due to power failures may be beyond one's control and, depending on the operation, a manufacturer may opt for backup power generators. Preventive equipment can be used to protect motors on an extrusion system due to power surges. Variations in power can affect the dimensions of the extrudate, resulting in rejects. In-house regrind facilities help to reduce material costs because these can be mixed with the 'virgin' material by ≤20% during an extrusion operation.

Flat or profiled extrudates coming out of an extrusion die should be calibrated appropriately and cooled sufficiently before reaching the take-up equipment to minimise rejects. In the case of electrical cables and pipes, for example, even thickness of walls must be enured. In modern machinery systems, this is done automatically by electronics but an operator must carry out checks periodically. Here, power variations and excess blockages in pre-die exit screens can cause problems due to excess foreign material in the resin polymer being used. If extra material is being used on an extrudate, the costs of the overall product will be affected. An operator must be vigilant and ensure that closed-loop auto-adjustment systems are working efficiently at all times. Sometimes, the start-up and purged material is too big (lumps) and hard for immediate re-use. This material can be ground separately using powerful shredders and sold or used for other products, thereby providing a kickback towards lowering costs.

Bibliography

1. http://www.britannica.com/science/polymer.
2. http://scifun.chem.wisc.edu/chemweek/polymers/polymers.html.

Reference

1. C. Defonseka in *Practical Guide to Flexible Polyurethane Foams*, Smithers Rapra, Shawbury, UK, 2013, p.93.

2 Cellular polymers

'Cellular foams', also known as 'foamed polymers', 'polymeric foams' or 'expanded plastics', have been important to human life since primitive people began to use wood (a cellular form of the polymer cellulose). Cellulose is the most abundant of all naturally occurring organic compounds, comprising about one-third of all vegetable matter in the world. Its name is derived from the Latin word *cellula*, meaning 'very small cells' or 'very small rooms'. The high strength-to-weight ratio of wood as well as the good insulating properties of cork, straw and other similar materials has contributed significantly as an incentive for research and development of a broad range of cellular synthetic polymers. The first commercial cellular polymer from natural sources was sponge rubber, whereas the first cellular synthetic polymer was an unpopular cellular phenol formaldehyde resin. Most plastics polymers can be foamed but only a relative few have found commercial use. Since 1940, cellular polymers have been in demand commercially in an increasing range of applications.

Of the widening range of cellular polymers being produced, the most popular and essential ones are cellular polyurethanes (PU) (for comfort), expanded polystyrene (EPS; for packaging and insulation) and cellular polyvinyl chloride (PVC; for floating devices, artificial leather and sheeting).

2.1 Need for cellular polymers

From ancient times, wood (which was available in abundance) was used for people's needs. Combined with metals such as copper, iron and alloys, people took important steps towards the needs and comfort of their daily lives. Although unaware of its cellular nature, they found other materials, for example, balsa wood, which was ideal for floating devices. People were looking to improve their comfort, clothing, and travelling modes, and relied heavily on naturally occurring sources. The discovery of crude oil and its vast potential made a very significant change and, as synthetic plastics polymers emerged, they replaced most traditional materials. Moreover, they were less costly, lasted longer, and could be produced in aesthetically pleasing colours and shapes. Although the initial plastics resins were solids or liquids, research showed the possibilities of converting/producing these resins in other useful forms.

Rapid expansion of air travel, automobiles, building-construction materials, packaging, consumer items, fishing industries, and bedding have come with ever-increasing demand due to global population growth. However, these necessities are associated with rising costs. For decades, cellular polymers have had a significant role in reducing costs, and their role will make more significant contributions with environmentally friendly water-blown expansion systems for cellular polymers.

https://doi.org/10.1515/9783110643121-002

2.2 Chemistry of cellular polymers

2.2.1 Classification

A 'cellular plastic' can be defined as a plastic made from a polymer of which the apparent density is decreased substantially by creation of numerous small cells throughout its mass. The polymer can be from a synthetic or natural source. Classification of cellular polymers has been made according to the method of manufacture, cell structure, or a combination of both.

The gas phase or expansion phase in a cellular polymer produces voids, pores or pockets called 'cells'. If these cells are interconnected in such a manner that gas can pass from one to another and the material has open cells, they may be classified as 'flexible foams'. If the cells are discrete and the gas phase of each is independent of that of the other cells and the material has closed-cells, they may be classified as 'rigid foams'. One test procedure from the American Society for Testing and Materials (ASTM) has suggested that foamed plastics can be classified as 'rigid' or 'flexible'. Using this test procedure, a flexible foam is one that does not rupture when a 20 × 2.5 × 2.5 cm is wrapped around a 2.5-cm mandrel at a uniform rate of 1 lap per 5 s at 15–25 °C; rigid foams are those that rupture under this same test procedure.

A 'structural foam' can be defined as flexible or rigid foams produced at densities >320 kg/m^3 and having cells in a foamed core rather than a typical lower-density foam structure of pentagonal-dodecahedron type. 'Integral foams' are also structural foams having a foamed core that decreases gradually in void content to form solid skins.

2.2.2 Theory of the expansion process

Cellular plastics can be prepared by several methods. The most important process consists of expanding a fluid polymer phase to a low-density cellular state which, in essence, is the expanding or foaming process. Other methods of producing a cellular state include leaching out solid or liquid materials that have been dispersed in a polymer, sintering small particles, and dispersing small cellular particles in a polymer.

The expansion process consists of three stages: i) creating small discontinuities or cells in a fluid or plastic phase; ii) causing these cells to grow to a desired volume; and iii) stabilising this cellular structure by physical or chemical means. 'Initiation' or 'nucleation' of cells is the formation of cells of such size that they can grow under the given conditions of foam expansion. Growth of a cell in a fluid medium at equilibrium is controlled by the pressure difference between the inside and outside of the cell, interfacial surface tension, and the radius of the cell. The pressure outside the cell is the pressure imposed on the fluid surface by its sur-roundings. The pressure inside the cell is the pressure generated by the blowing agent dispersed or dissolved in the fluid. If blowing pressures are low, the radii of initiating cells must be large. The hole that acts as an initiating site can be filled with a gas or a solid that breaks the fluid surface and thus enables the blowing agent to surround it.

During the time of cell growth in a foam, several properties of a system can change considerably. Cell growth can, therefore, be treated only qualitatively. There are four considerations of primary importance. The first consideration is that the fluid viscosity will change considerably, influencing the rate of cell growth and flow of polymer to intersections from the cell walls, leading to collapse. Second, the pressure of the blowing agent decreases, falling off less rapidly than an inverse-volume rela-tionship because new blowing agent diffuses into the cells as the pressure falls. Third, the rate of growth of the cells is dependent upon the: viscoelastic nature of the poly-mer phase; pressure of the blowing agent; external pressure on the foam; permeation rate of blowing agent through the polymer phase. The final consideration is that the pressure in a cell of a smaller radius is greater than that in a cell of larger radius. Thus, there is a tendency for pressures to equalise by breaking the walls separating the cells or by diffusion or control of the blowing agent from smaller to larger cells.

Stabilisation of a cellular state is very important. The increase in surface area corresponding to formation of many cells in the plastics phase is accompanied by a release and increase of free energy from the foaming system. Methods of stabilisa-tion of foamed systems can be chemical or physical.

- *Chemical stabilisation*: The chemistry of the system used determines the rate at which the polymer phase is formed and the rate at which it changes from a vis-cous fluid to a dimensionally stable crosslinked polymer phase. It also governs the rate at which the blowing agent is activated, whether it is due to rise in tem-perature or to insolubility in the liquid phase. The type and amount of blowing agent used governs the: amount of gas generated; rate of generation; pressure that can be developed to expand the polymer phase; amount of gas lost from the system relative to the amount retained in cells.

  Additives to a foaming system (agents that control cell growth) can influ-ence nucleation of foam cells considerably through their effect on the surface tension of the system or by acting as nucleating sites from which cells can grow. They can also influence the mechanical stability of the final structure of solid foam considerably by changing the physical properties of the plastic phase, which allows the blowing agent to diffuse from cells to the surround-ings. Environmental factors such as temperature and pressure also influence the behaviour of thermoset foaming systems.

- *Physical stabilisation*: In physically stabilised foaming systems, the factors are es-sentially the same as for chemically stabilised systems but for different reasons. The chemical composition of a polymer phase determines the temperature at which foam can be produced, the type of blowing agent required and the cooling rate of the foam necessary for dimensional stabilisation. The composition and concentration of blowing agents control the rate at which gas is released, the pressure generated by the gas, and the escape or retention of gas from the foam cells for a given polymer after the blowing agent has been activated.

Additives have the same effect on thermoplastic foaming processes as on thermoset foaming processes. Environmental conditions are important in this case because of the necessity of removing heat from the foamed structure to stabilise it. For the same reason, the dimensions and size of the foamed structure are also important.

2.3 Types of cellular polymers

2.3.1 Expanded polystyrene

EPS raw materials are available as tiny round beads (similar to grains of sugar), colourless or coloured, with each bead containing a small amount of an inert gas (e.g., pentane) which, acting as a blowing agent, expands under heat to form cellular foams. These beads can be expanded up to 40–50-fold their original size, thus creating an array of densities. There are many grades based on bead sizes made especially from general purpose to insulation to marine applications. Standard grades containing petroleum-based blowing agents are called EPS, whereas the water-blown ones are identified as water-blown expandable polystyrene (WEPS).

2.3.2 Cellular polyurethanes

Cellular PU can be categorised broadly into flexible polyurethane foam (PUF) and rigid PUF. The former have open-cells, whereas the latter have closed cells and, in most cases, a protective skin. Cellular PUF are, in general, blown with water as the blowing agent to form cells from solid polyols in liquid form. Some foam manufacturers may use an additional blowing agent such as methylene chloride if very low densities are desired. Petroleum-based polyols do not present expansion problems with water, but some of the agricultural-based polyols from vegetable-oil sources may present problems with cell formation.

2.3.3 Cellular polyvinyl chloride

PVC is a very versatile polymer available in forms such as powders, pellets or other forms. Cellular PVC can be produced from several expandable formulations as well as decompression methods. Flexible or rigid foams can be made depending on the amounts and types of blowing agents and plasticisers used. Latest technologies involve using PVC for production of polymeric composites, and cellular PVC enhances weight reduction in composites.

2.3.4 Polyethylene cellular foams

Cellular PE foams are very versatile and useful materials in many industrial and domestic activities. Their cost-effectiveness plays a big part in many spheres of activity. These foams are available in many colours and shapes as well as open- or closed-cell foams. PE has a sharp melting point and its viscosity decreases rapidly over a narrow temperature range above the melting point. Hence, production of a low-density polyethylene (LDPE) foam with nitrogen or chemical blowing agents (CBA) is difficult because the foam collapses before it can be stabilised. This problem can be eliminated by crosslinking the resin before it is foamed, which slows the viscosity decrease above the melting point and allows the foam to be cooled without collapse of the cell structure.

2.3.5 Polyisocyanurate cellular foams

The isocyanurate ring formed by the trimerisation of isocyanurates is known to possess high resistance to heat and flames, as well as low generation of smoke during burning. Crosslinking *via* the high functionality of the isocyanurate system produces a foam with inherent friability. Modification of the isocyanurate system with a longer-chain structure, such as that of polyether polyols or terephthalate-based polyester polyols, increases the abrasion resistance of the resultant foam.

2.3.6 Polyphenol cellular foams

Polyphenol cellular foams are thermoset foams. An important example of a chemical-stabilisation process is production of phenolic foams by crosslinking polyphenols. A typical phenolic foam system consists of liquid phenolic resin, blowing agent, catalyst, surface-active agent and modifiers. Various formulations can be used to improve one or more properties of the foam for specific applications. Principal features of phenolic foams are low flammability, solvent resistance and excellent dimensional stability over a wide temperature range, thus making them good thermal-insulating materials.

2.3.7 Epoxy, silicones and polyimide cellular foams

Cellular foams can also be formed from epoxy, silicone and polyamide resins by a chemical stabilisation process. If specific properties are essential (e.g., in applications such as for aircraft, ships and railways), these specialty foams play an important part. These foams can be used for low generation of smoke, high temperature resistance, fire resistance and chemical resistance. Several processes can be used to produce phenolic foams, such as continuous production of free-rising foam for slabs and slabstock (similar to that for PUF), foam-in-place batch process, sandwich panelling and spraying.

2.3.8 Cellular cellulose acetate

An extrusion process can also be used to produce cellulose acetate foams in the density range 96–112 kg/m^3. A hot mixture of polymer, blowing agent and nucleating agent is forced through an orifice into the atmosphere onto a conveying device. The hot melt expands, cools and is carried away on the conveyor. In general, cellulose-acetate foams exhibit a coarse morphology with non-homogenous distribution of cells. The relatively low reduction in density and coarse morphology of the foam with only a few but large cells is typical for foams produced with CBA. In comparison with an extruded polystyrene (XPS) foam board produced with physical blowing agents (PBA), cellulose-acetate foams are stiff and have a high tensile modulus due to the relatively high amount of compact matrix material around the bubbles, which determine the mechanical properties.

The rigidity in combination with high heat resistance and ability of these cellulose-acetate foams for thermoforming makes them attractive for applications using rigid foams. Moreover, the excellent injection-moulding possibilities of these foams make them ideal for compact parts with foam cores. Ongoing research has elicited methods for making these foams with PBA that can result in more homogenous cells and finer foam morphologies. To achieve fine low-density foams produced with PBA, the polymer properties have to fulfil specific requirements that may require the cellulose acetate to be modified in terms of rheological, thermal and physical properties.

2.3.9 Thermoplastic structural foams

Structural foams having integral skins with cellular cores and high strength-to-weight ratios are formed by injection moulding, extrusion or casting depending on product requirements. The two injection-moulding processes used most widely are low-pressure and high-pressure processes. In the low-pressure method, a shot of resin containing a blowing agent is injected into a mould, in which the material expands to fill the mould under pressure. This process produces structural foam products with a characteristics surface 'swirl' pattern produced by the collapse of cells on the surface of moulded articles.

In the high-pressure process, a resin melt containing a CBA is injected into an expandable mould under high pressure. Foaming begins as the mould cavity expands. This process produces structural foams with very smooth surfaces because the skin is formed before expansion occurs.

Conventional extruders with specially designed dies can be used to produce extruded structural foams. Large structural foam products are produced by casting expandable plastic pellets containing a blowing agent into aluminium moulds on a chain conveyor. The tightly clamped moulds go through a heating zone, in which the pellets soften, expand and fuse together to form cellular products. Then, moulds pass through a cooling zone. This process produced foam products with uniform closed-cell structures but without solid skins.

2.3.10 Thermoplastic microcellular foams

Microcellular foams are foam structures formulated and produced to yield 'micropores' or 'microcells' (cells much smaller than cells in standard cellular foams in the polymer matrix). These microcellular foams can be divided further into two classes, microcellular and ultra-microcellular, with the difference being in the size of cells. These cells are very small in diameter so, to a casual observer, these foams appear as solid plastics. Advantages of microcellular plastics are: reduced consumption of material; accurate parts; long-term stability due to low residual stresses in moulded parts; higher productivity due to shorter cycle times; having light weight and unique appearances. The tight and very small structures as well as uniform spread yield superior mechanical properties compared with conventional foams. According to recent reports, production of 'nanocellular' foams is being researched.

2.3.11 Syntactic cellular polymers

Syntactic cellular polymers are produced by dispersing rigid, foamed, microscopic particles in fluid polymer and then stabilising the system. The particles are, in general, spheres or 'micro-balloons' of phenolic resins, urea formaldehyde (UF) resins, or copolymers of vinylidene chloride, acrylonitrile, glass or silica with diameters of 30–120 µm. Commercial micro-balloons have densities of ≈144 kg/m^3. The fluid polymers used can be the usual coating resins (e.g., epoxy resins, polyesters and UF).

The resin, catalyst and micro-balloons are mixed to form a mortar, which is then cast into a desirable shape and cured. Very specialised electrical mechanical properties can be obtained by this method but at higher cost. This method of producing cellular polymers is applicable for small-quantity specialised applications, and one advantage is that it requires very little specialised equipment. Foams can also be produced using other methods.

2.3.12 Microcellular open-cell low-density polyethylene foam sheets

Most open-cell cellular foams used to be manufactured industrially with PU thermoset polymers but, as technology advanced, thermoplastic polymers began to be used. A wide range of conventional polymers, such as polyethylene (PE), polypropylene, polyethylene terephthalate, PVC, and polystyrene (PS), is now available for extrusion foaming. One method for extrusion of LDPE/PS blends uses carbon dioxide (CO_2) and butane gas as blowing agents and a small amount of crosslinking agents to achieve highly open-cell filament foams. However, these materials are used primarily for applications such as filters, separation membranes, sound-insulation panels, and battery separators, so the requirement is for more industry-friendly sheet foams rather than filament foams.

A single-screw foaming extrusion system can be used in this manufacturing process. The main function of the screw is for mixing of the polymer resin into a homogenous hot melt. The gas-injection equipment attached to the extruder is used for injecting the soluble amount of gases into the hot melt. The gear pump controls the flow rate of the polymer melt, whereas a dissolution-enhancing device ensures the homogeneity of the polymer/blowing-agent solution. Heat exchangers provide uniform cooling for the polymer melt. The annular die used provides the shape and nucleation of the extrudate.

A general method for extrusion of foam sheets for LDPE or LDPE/PS is for these pellets to be mixed with 2.0 wt% talc and a small amount of a crosslinking agent and fed into a hopper of a conventional plastics extruder. A barrel consisting of three or four zonal heating sections and the rotating motion of the screw produce a hot melt.

Then, calculated amounts of CO_2 and butane are injected into this hot melt at two ports by pumps. They blend to form one homogenous mass and are then fed into a heat exchanger, where the hot mass is cooled to a pre-selected temperature. Then, the cooled polymer/gas solution enters a die of pre-determined shape and size, and foaming occurs as the extrudate exits into the atmosphere. Different qualities of foam cells, sheet widths, sheet thicknesses and other required properties can be achieved by variation of the important parameters of this extrusion system: screw speed; ratio of screw length:diameter; speed of the gear pump; barrel temperature; contents of the blowing agent. The extrudate will be white or colourless, so colouring dyes or colour masterbatches can be added to the polymer at the start to achieve the desired colour, in addition to the standard additives that will be also mixed into the polymer.

2.4 Cellular polymers with greatest application

2.4.1 Expanded polystyrene

Cellular PS can be categorised broadly as EPS, XPS and WEPS.

In general, EPS comes in the form of very small beads (with each bead containing a tiny amount of a blowing agent) and is colourless. Blowing agents used to be petroleum-based and emitted harmful gases to the atmosphere. Subsequently, they were banned, and now an inert gas (e.g., pentane) is used widely, which hardly effects the environment. Many grades are manufactured by chemical companies (e.g., BASF, Dow Chemicals, and Bayer AG) suitable for a wide range of applications. Coloured grades are also available but the colours must be incorporated during polymerisation for effective colouration. The products from EPS are white and, if colouring is done after pre-expansion, the result would be a 'mottled' surface colour.

Grades of EPS are made specially for building construction, with fire-retardant grades being the most popular. Generally, they come in three different grades of properties such as general insulation, moisture resistant and fire retardant; and are coloured for easy identification. These insulation sheets can be cut from large moulded blocks or extruded in sheet form. The building-construction industry prefers extruded sheets because they have a protective skin on both sides, thus giving additional insulation.

BASF has produced an environmentally superior EPS from a new polymer called Neopor® for the building-construction industry. It provides higher thermal insulation without using hydrofluorocarbon blowing agents, which may contribute to global warming. This unique material contains tiny amounts of graphite within the polymer matrix of EPS, and the finished boards have a grey–platinum colour. The graphite particles reflect and absorb radiant energy, thereby increasing the insulation capacity (R-value) by ≤20% with reduction of the transmission of radiant energy within the insulation itself. Neopor® is used worldwide in insulation applications in which cost-effectiveness and sustainability are priorities.

Standard grades of EPS can be moulded into any shape or size. Hence, moulders create various products for more effective insulation, cost-effectiveness, and environmental friendliness. One such system, Insulating Concrete Forms (Plasti-Fab Limited) is used widely in large and small residential projects as a solution to wall-assembly needs. These insulating forms come in various types and sizes. They conform to building regulations and have studding between the slabs of foam. In this way, concrete can be poured into the voids to give additional structural strength with superior barriers to vapour and air and form walls automatically, thus saving construction time. This concept of two layers of EPS insulation with web connectors moulded into the EPS insulation, and the top and bottom with pre-formed interlocking mechanisms to provide accurate alignment, enable easy and quick surface finishes.

XPS boards are composed of small, tightly packed closed cells that entrap a low thermally conducting gas, which contributes to the insulating effect. This characteristic also provides these insulation boards with water-impermeability properties. Some manufacturers have a colour code: pink, blue and green to indicate the different R-values and other special properties. XPS boards have a smooth skin on both sides that acts as an added vapour/air barrier and gives higher compressive strength, thereby eliciting better insulation properties than EPS board cut from large moulded blocks.

The extrusion process for XPS boards is in several steps. Briefly, extrusion-grade PS pellets and additives are introduced into a hopper of a regular tandem extrusion system and melted by heat. A foaming agent (usually CO_2) is injected into this hot melt and homogenised. As this hot melt is metered out through a T-die of predetermined size, foaming takes place as the hot melt exits through the T-die due to change in pressure as it enters the atmosphere. Then, the foamed extrudate is controlled for size and shape between two calibrating plates. A set of rollers after the plates also helps to control foam growth and convey the foamed sheet forward, where it is cooled and cut to desired lengths. Pre-set quality-control instruments ensure exact measurements, the desired thickness, and that the foamed boards are cooled and allowed to mature before final finishing. A printing process enhances technical, manufacturer identity, and other information for marketing purposes.

The EPS raw material is supplied by manufacturers in 25-kg paper bags, 200-kg steel drums, or 400-kg bulk packs ('gaylords'). Grades range from general-purpose to marine applications.

2.4.2 Cellular polyurethanes

Flexible PUF and rigid cellular PUF offer vast possibilities in the domestic, industrial, building construction, automobile, and aircraft sectors. Moreover, these products have been replacing traditional materials gradually, and proving to be an essential part in daily life.

The two basic components in a PU production formula are a polyol and an isocyanate, both of which are in liquid form. To achieve good results, the recipe must be formulated carefully. A typical base formulation comprises fillers, stabilisers, catalysts, blowing agents, pigments, surfactants, and water in addition to the polyol and isocyanate. Water acts as the blowing agent and, if very low densities are desired, an additional supportive blowing agent such as methylene chloride can be incorporated. Densities can range from very low, flexible, to rigid foams. This is a highly flammable heat-giving (exothermic) reaction process, so appropriate safety precautions are advised. The chemicals (especially the polyol and isocyanate) must be handled carefully because they are harmful.

The urethane-forming components are the polyols and isocyanates. Water reacts with the isocyanate in the formula to generate CO_2 and urea groups that modify the polymeric structure. This vigorous reaction is the prime source of exothermic heat that triggers foaming. The CO_2-forming reaction is known as the 'blowing reaction' and the urethane-forming reaction as the 'gelling reaction', which is the primary means of polymerising the starting materials into a long-chain polymer network. The amount of water present controls the blowing range, whereas the rates of blowing and gelling will be controlled by the catalysts chosen. In general, tertiary amines are used to control the blowing reaction and organometallics are used to promote gelation though, in practice, both contribute to both reactions. Silicone surfactants are used to control the size and uniformity of cells through reduced surface tension and, in some cases, to assist in solubilisation of the various reactants. For rigid PUF, lower-cost non-silicone surfactants based on copolymers of butylene oxide and ethylene oxide can be used.

Viscoelastic foams or PUF known as 'memory foam' are specially formulated foams to increase density and viscosity. They are soft foams with low resilience and pressure-sensitive properties. Unlike standard flexible PUF, these foams are four-dimensional and provide maximum comfort by quickly conforming and moulding itself to the shape of the body. That is, mattresses or mattress toppers made from this material 'accept' a body triggered by the temperature of a body and its weight, thus providing extra comfort and also providing the sleeper with a sensation of 'floating on a cloud'. If the body is removed, the foam reverts to its original state.

Viscoelastic foams are open-cell structures and formulated with special polyols that are based on petroleum or agriculture (e.g., vegetable oils). Their formulation is slightly more complicated than that for standard PUF. Also, incorporation of additives can achieve the desired qualities which, in general, are graded by their densities. Additives also protect the foam and dyes/pigments can produce foams of pleasing colours.

Applications for PUF are mattresses, cushions, sheets, slabs, carpet underlay, footwear, medical applications, space-travel comfort, aircraft seats, automobile seats, mattress toppers, sleeping bags, domestic/industrial sponges, bicycle seats, furniture, upholstery, adhesives, as well as rebounded material for mattress pads and pillows.

2.4.3 Cellular polyvinyl chloride

Cellular PVC can be produced by many methods, some of which use decompression processes. Unlike the typical process used for thermoplastic resin foams, in which the melt is heated to a temperature considerably above its second-order transition temperature so that the resin can flow, PVC requires the assistance of a plasticiser to fuse into a plastisol resin. This process is used because PVC resin is susceptible to thermal degradation.

Fusion of a dispersion of PVC resin in a plasticiser provides a unique type of physical stabilisation. The viscosity of a resin–plasticiser dispersion shows a sharp increase at the fusion temperature. In such a system, expansion can take place at a temperature corresponding to the low viscosity, and then the temperature can be raised to increase viscosity and stabilise the expanded state.

Extrusion processes can also be used to produce high- and low-density flexible cellular PVC. Usually, a decomposable blowing agent is blended with the compound before extrusion. Then, this compounded resin batch is fed into an extruder, where it is melted under pressure and metered out of an orifice into the atmosphere. The extrudate is cooled and hauled off on a conveyor. Another extrusion process involves pressurisation of a fluid plastisol at low temperatures with an inert gas. Subsequently, this mixture is extruded onto a conveyor belt or into moulds, where it expands. Then, the expanded dispersion is heated to fuse into a dimensionally stable form. Injection moulding of high-density cellular PVC products can be accomplished by injection of the plastisol into moulds and cooling within them before removal.

Applications for cellular PVC products include packaging, insulation, sound dampening, floating devices, sleeping bags, artificial leather, sports mats, footwear, handbags, clothing accessories, floor mats, sheeting, electrical cable sheaths, floor tiles, sportswear, and padding.

Bibliography

1. G. Abhishek, *Materials Letters*, 2013, **94**, 76.
2. G. Abhishek, *Journal of Applied Polymer Science*, 2014, **18**, 131.
3. S.T. Lee in *Foam Extrusion: Principles and Practice*, 2nd Edition, CRC Press, Baco Raton, FL, USA, 2014.
4. L.B. Bottenbruch in *Technische Thermoplaste: Polycarbonate, Polyacetale, Polyester, Celluloseester: Kunststoff-Handbuch 3/1, Technische Thermoplaste*, Hanser Fachbuchverlag, Germany, 1992 [In German].
5. J.E. Mark in *Polymer Data Handbook*, Oxford University Press, New York, NY, USA, 2009.
6. P.C. Lee, J. Wang and C.B. Park, *Journal of Applied Polymer Science*, 2006, **102**, 4, 3376.

3 Water-blown cellular polymers

3.1 Chemistry of water

In the context of water blowing of polymers to produce cellular materials, under-standing the chemistry of water is important. Water is the most abundant molecule on the Earth's surface and one of the most interesting molecules to study. A molecule of water consists of one atom of oxygen bound to two atoms of hydrogen. Some other names for water are dihydrogen monoxide, oxidane, hydroxylic acid and hydrogen hydroxide.

The hydrogen atoms are bonded to one side of the oxygen atom, resulting in one molecule of water having a positive charge on the side where the hydrogen atoms are and a negative charge on the other side. Opposite electrical charges attract, so water molecules tend to attract each other, making water 'sticky'. The side with the hydro-gen atoms (positive charge) attracts the oxygen side (negative charge) of a different water molecule, so they tend to clump together. Water is called the 'universal solvent' because it dissolves more substances than any other liquid. Pure liquid water at room temperature is odourless, tasteless and nearly colourless. Water has a very faint blue colour, which becomes more apparent in large volumes of water. Pure water has a neutral pH value (level of acidity/alkalinity) of ≈ 7, which means it is neither acidic nor basic.

3.1.1 Physical properties of water

Water is unique in that it is the only natural substance found in all three states, liquid, solid (ice) and gas (steam), at temperatures found on Earth, where water is constantly interacting, changing and in movement. Water freezes at 0 °C and boils at 100 °C. In general, the freezing point and boiling point of water are the 'baseline' with which temperature is measured. Water is unusual in that its solid form (ice) is less dense than its liquid form, which is why ice floats. Water has a high specific heat index, so it can absorb a lot of heat before it begins to get hot. The high specific heat index of water helps regulate the rate at which air changes temperature and is particularly helpful to industry (e.g., water used as a coolant in the radiator of a car). Water also has high surface tension, which is why it is sticky and elastic and tends to clump together as drops rather than spread out in a thin film. Surface tension is responsible for capillary action, which allows water (and its dissolved substances) to move through even matter of very small-cross section.

https://doi.org/10.1515/9783110643121-003

3.1.2 Water temperature

Water temperature is useful in daily life, to industries, and the oceans (for the survival of marine life). A lot of water is used for cooling many industrial operations where, in general, cool water is used initially and warmer water is released into the environment. The temperature of water if it is used as a coolant in processing of plastic polymers is important and, if it is recirculated in a system, it should be cooled before re-entering cooling areas. If water is used as a blowing agent in polymers, the temperature and quality of the water used becomes important if maximum results are to be obtained.

3.1.3 pH value of water

The pH value of water is a measure of how acidic/basic it is. The range is 0–14, with 7 being neutral. pH values <7 indicate acidity, whereas a pH value >7 indicates a base. pH is a measure of the relative amount of free hydrogen and hydroxyl ions in the water. Water that has more free hydrogen ions is acidic, whereas water having more free hydroxyl ions is basic. pH can be affected by chemicals in the water, so the pH value is an important indicator of water that is changing chemically. This is an important factor if water is formulated as a blowing agent in polymers to produce cellular structures. pH is reported in logarithmic units, just like the Richter scale, which is used to measure the magnitude of earthquakes. Each number represents a tenfold change in the acidity/basicity of the water. For example, water with a pH of 5 is tenfold more acidic than water having a pH of 6.

3.1.4 Specific conductance

Specific conductance is a measure of the ability of water to conduct an electrical current. It is highly dependent on the amount of dissolved solids (e.g., salt) in the water. Pure water, such as distilled water, has very low specific conductance and seawater has high conductance. Rainwater often dissolves airborne gasses and airborne dust while it is in the air, so often has higher specific conductance than distilled water. Specific conductance is an important measurement of water quality because it gives a good indication of the amount of dissolved material in the water. This is also a factor if using water as a blowing agent for polymers (though electrical conductance may not be important).

3.1.5 Hardness

'Hardness' is an important factor if water is to be used as a blowing agent for polymers. The amount of dissolved calcium and magnesium in water determines its hardness. Water hardness will vary from area to area. If the water is relatively hard, it is difficult to lather when washing hands or clothes. If hard water is the main source for industries and manufacturing processes, chemicals must be used to 'soften' the water before use because it will corrode and damage equipment in the long-term. For example, steam-generator tubes corrode fast if hard water is not treated before use.

3.2 Blowing agents

A blowing agent is a substance that can produce a cellular structure *via* a foaming process in various materials that undergo phase transition, such as polymers, plastics, metals, and ceramics. Blowing agents can take the form of liquid, powder, pellets or pastes. Typically, they are incorporated if the material to be 'blown' is in a liquid state or molten state (e.g., polymers). Main purposes of creating cellular structures are for cost savings and making the material light, but these cellular structures have other valuable end-uses: sound absorption; building insulation; integral skin foams for automobiles; geo-membranes; comfort items; elasticity; permeability; shock absorbance; electrical insulation; packaging.

Blowing agents (also known as 'pneumatogens') or related mechanisms to create voids or holes in a matrix producing cellular materials are classified as physical blowing agents (PBA), such as chlorofluorocarbons and hydrofluorocarbons (which have been banned because they deplete the Earth's ozone layer) and hydrocarbons [e.g., pentane and liquid carbon dioxide (CO_2)]. The bubble/foam-making process is endothermic (needs heat to be activated). Chemical blowing agents (CBA) such as azodicarbonamide, hydrazine and other nitrogen-based, methylene chlorides used for thermoplastic and elastomeric foams react chemically to form cellular structures. Blowing reactions produce low-molecular-weight compounds acting as the blowing gas, so additional exothermic heat (heat giving) is released. Powdered titanium hydride is used as a foaming agent in the production of metal foams because it decomposes to form hydrogen gas and titanium at elevated temperatures.

Mixed PBA/CBA systems are used to form foams with low densities such as flexible polyurethane foams (PUF), in which water is the primary blowing agent for standard low-density foams and water/methylene chloride combinations are used for very-low-density foams. Here, physical and chemical processes are used in tandem to balance each other out with respect to thermal energy released/absorbed, thus minimising a temperature rise. Otherwise, excessive loading of a PBA can cause excessive exothermic heat, which can cause thermal degradation of a developing thermoset or thermoplastic material. For example, to avoid this problem in polyurethane (PU) foaming systems (in which isocyanate and water react to form CO_2 in a gaseous reaction to cause foaming), addition of liquid CO_2 is helpful, especially at very low densities.

Emerging technologies are using water-blown systems for making cellular foams, especially for PU, polystyrenes (PS) and other polymers presented in subsequent chapters in this book.

3.2.1 Mechanically made foams

Mechanically made foams/froths involve the methods of producing bubbles in liquid polymer matrices (e.g., an unvulcanised elastomer in the form of latex). Methods may include: simple whisking in air or other gases; low-boiling volatile liquids in low-viscosity lattices; injection of a gas into a hot melt in an extruder or into injection-moulding barrels and allowing the shear/mix action of the screw to disperse the gas uniformly in a hot polymer melt to form very fine bubbles (cellular structures). If the melt is moulded or extruded and the part is at atmospheric pressure, the gas comes out of the melt and expands the polymer melt immediately before solidification. Frothing (akin to beating of egg whites) is also used to stabilise foamed liquid reactants (e.g., to prevent slumping occurring on vertical walls (insulation) before cure). That is, avoiding foam collapse and sliding down a vertical face due to gravity. Spray systems are available that do not need stirring and the foam stabilises immediately upon release.

Two practical applications are mentioned here. Soluble fillers (e.g., solid crystals of sodium chloride) are mixed into a liquid urethane system, which is shaped into a solid polymer part. Later, the sodium chloride is washed out by immersing the solid moulded part in water for some time. This strategy results in interconnected holes/cells in relatively high-density polymer products (e.g., synthetic leather material for shoes).

Hollow spheres and porous particles (e.g., glass shells, glass spheres, fly ash) are mixed and dispersed in liquid reactants, which are shaped into solid polymer parts containing a network of voids/cells.

3.3 Water-blown hybrid urethanes

A company called Natural Polymers has designed an exceptional PUF system that uses polymers from natural (non-petroleum) sources. Their polymer chemistry is different from other biological- and agricultural-based systems in that they use a proprietary process designed to maximise the polymers that nature has made available.

This company's water-blown hybrid urethanes have one of the highest bases of biological raw material in the polymer industry, with the lowest amount of embodied energy in the finished product. Their thinking and goals are to reduce the carbon footprint through innovative technology and designs through innovative chemistry.

Conventional polymer chemistry is based solely on propylene oxide and ethylene oxide, which are petroleum-based building blocks. Other companies have tried to develop biological- and agricultural-based foams by replacing small percentages of the petroleum content with vegetable-based polyols which, in many cases, depend heavily on petroleum-based raw materials. In most cases, the resulting finished products contain ≈90% petroleum-based content. The products made by Natural Polymers are different in that they can reduce the amount of petroleum content by ≤30% yet maintain the performance and physical properties of a 100%-based system.

3.4 End uses of water-blown foamed polymers

Cellular materials made from foam-blowing agents encompass a wide variety of applications: air-conditioning, trucks, automobiles, furniture, bedding, and packaging. Some of the most popular end-uses are:

- Rigid PUF: appliance foams include insulation foam in domestic refrigerators and freezers.
- Rigid PU spray foams: commercial refrigeration, sandwich panels, insulation for roofing, walls, pipes, metal doors, vending machines, coolers, buoyancy and refrigerated trucks.
- Flexible PUF: furniture, bedding, chair cushions, shoe soles, and sponges.
- Integral skin PUF: steering wheels, dashboards and bumpers of cars, and shoe soles.
- Expandable polystyrene (EPS) foams: packaging, building insulation, floatation devices, hot/cold packs, road-construction specialties, pipe insulation, adhesives.
- Foamed polyolefins: sheets and tubes for packaging and pipe insulation.
- EPS foam: extruded sheets building insulation for walls, roofing and floors.
- Polyisocyanurate foams: spray foams for insulation of roofs and walls.
- Polyvinyl chloride (PVC) foams: shoe soles, protective wear, flotation devices, artificial leather.

3.5 Chemical properties

Polymers are substances whose molecules have high molar masses and are composed of a large number of repeating units. There are naturally occurring polymers and synthetic polymers. Among naturally occurring polymers are proteins, starches, cellulose and latex. Synthetic polymers are produced commercially mainly from the byproducts of crude oil, vegetable oils and agricultural-based products.

Polymers are formed by chemical reactions in which a large number of molecules called 'monomers' are joined together to form a chain. In many polymers, only one monomer is used. In other polymers, two or three different monomers may be combined. Polymers are, in general, classified by the characteristics of the reactions by which they are formed. If all atoms in the monomer are incorporated into the polymer, the polymer is called an 'addition polymer'. If some of the atoms of the monomers are released into smaller molecules, such as water, the polymer is called a 'condensation polymer'. Most addition polymers are made from monomers containing a double bond between carbon atoms. Such monomers are called 'olefins' and most commercial addition polymers are 'polyolefins'. Condensation polymers are made from monomers that have two different groups of atoms that can join together to form, for example, an ester or amide links. Polyesters are an important class of commercial polymers, as are polyamides (PA) (e.g., Nylon).

3.5.1 Polyethylene terephthalate

Polyethylene terephthalate or polyethylene terephthalic ester (PETE) is a condensation polymer produced from the monomers, ethyl glycol, a di-alcohol and dimethyl terephthalate. By transesterification, these monomers form ester linkages between them to yield a polyester. PETE fibres are manufactured under different trade names such as Dacron™ and Fortrel™. These fibres are ideal for making synthetic fabrics, especially if non-ironing properties are important (e.g., 'permanent press' fabrics). Mylar™ is a trade name for PETE film, and this polymer can be made into transparent sheets and castings. Another application is making transparent bottles for carbonated beverages.

3.5.2 Polyethylene

Polyethylene (PE) is probably the simplest and most popular polymer in the market and has a wide variety of applications. PE polymers are composed of chains of repeating carbon/hydrogen units. They are produced by the addition polymerisation of ethylene – a byproduct of petroleum. PE properties depend on the manner in which ethylene is polymerised. If ethylene is catalysed by organometallic compounds at moderate pressure (15–30 atmospheres), the product is high-density polyethylene (HDPE). Under these conditions, the polymer chains grow to very great lengths and molar masses average many hundred-thousands. HDPE is hard, tough and resilient. If ethylene is polymerised at high pressure (1,000–2,000 atmospheres) at elevated temperatures (190.5–210 °C) and catalysed by peroxides, the product is low-density polyethylene (LDPE). This form of PE has a molar mass of 20,000–40,000 g. LDPE is relatively soft and two of the main applications are plastic films and containers.

3.5.3 Polyvinyl chloride

Polymerisation of vinyl chloride produces a polymer similar to PE but having chlorine atoms at alternate carbon atoms on the chain. PVC is rigid and brittle. About two-thirds of PVC produced annually is used in pipe manufacture. It is also used in the production of vinyl siding for houses and clear plastic bottles. If PVC is blended with a plasticiser, such as a phthalate ester, it becomes pliable and is used to form flexible articles such as raincoats, shower curtains and for coating on fabrics to make artificial leather.

3.5.4 Polypropylene

Polypropylene (PP) is produced by the addition polymerisation of propylene. Its molecular structure is similar to that of PE except that it has a methyl group ($-CH_3$) on alternate carbon atoms of the chain. Its molar masses are very high. PP is slightly more brittle than PE but softens at a temperature 40 °C higher. PP is used extensively in the automotive industry for interior trim, instrument panels and food packaging. It is an ideal matrix for the production of polymeric composite resins consisting of PP and biomass reinforcements such as wood flour, rice hulls, and coir dust. It is formed into fibres of very low absorbance and high stain resistance to be used in clothing, home furnishings and carpeting. Its fibres are used widely as an alternative to grass on playing fields.

3.5.5 Polystyrene

Styrene monomers polymerise readily to form PS, a hard, highly transparent polymer. The molecular structure is similar to that of PP but with the methyl groups of PP replaced by phenyl groups ($-C_6H_5$). A large portion of this polymer (especially foamed material) goes into packaging and building insulation. PS is also used widely for food packaging and containers. PS is foamed readily by incorporation of a tiny amount of an inert gas (e.g., pentane) and now water-blown PS beads (which are more environmentally friendly) are being used. Styrofoam is a well-known brand name for EPS. If rubber is dissolved in styrene before polymerisation, the PS produced is much more resistant to impact. This type of PS is used extensively in home appliances, refrigerator bodies, and housing for air conditioners.

3.5.6 Polytetrafluoroethylene

Teflon™ is a well-known brand name and is made of polytetrafluoroethylene (PTFE). It is formed by the addition polymerisation of tetrafluoroethylene. PTFE is distinguished by its complete resistance to virtually all chemicals and by its silky slippery surface. It maintains its physical properties over a large temperature range (371.7–385 °C). These properties make it especially useful for components that must operate under harsh chemical conditions and extreme temperatures. Its most familiar household use is as coating on cooking utensils (non-stick).

3.5.7 Polyurethane

PU is an important and versatile class of polymers formed by the addition polymerisation of a diisocyanate (whose molecules contain two -NCO groups) and a dialcohol (two -OH groups). The polymer chain is linked by urethane groups (-O-CO-NH-). The –NH- portion of the urethane group can react similarly with an –OH group to produce crosslinking between polymer chains. PU can be spun into elastic fibres called Spandex™ and Lycra™.

Some of the biggest applications of foamed PU are for bedding, furniture and upholstery, and carpet underlay. Hard foams also have a wide variety of applications: automobiles, integral foams and shoe soles. Open- (soft) or closed (rigid)-cellular structures are obtained by foaming of the polyol polymer by CO_2 generation from the reaction of water with an isocyanate. This is an exothermic reaction, so various additives are used to control foaming and cell formation to obtain uniform cell structures. This foaming process is difficult and the components must be formulated carefully. Safety is a priority during foaming due to corrosive materials and the possibility of fire. Water is the primary blowing/foaming source in this reaction, but methylene chloride is also used if very low foam densities are desired.

3.5.8 Polyamide

PA are a group of condensation polymers commonly known as Nylons. Nylon is made from two monomers: one dichloride and one diamine. One particular Nylon is called Nylon-6,10 because it contains alternating chains of 6 and 10 carbon atoms between nitrogen atoms. Nylon is a very versatile industrial polymer and be formed readily into fibres that are strong and long-lasting, making them well suited for use in carpeting, upholstery fabric, tyre cords, brushes, and turf for playing fields. Nylon is also formed into rods, bars and solids that can be machined easily. Nylon is an ideal substitute for gear wheels and similar products as an alternative to metal.

3.6 Physical properties

3.6.1 Density

The first important physical property is the density (measured in kg/m^3) of the foamed material. For example, if a cube of foam measures $1 \times 1 \times 1$ m and weighs 2 kg, the density is 2 kg/m^3. Density is related directly to foam cost: the higher the density, the costlier the foam. Densities can range from very soft to hard foams, and the selection depends on the end-application. Manufacturers of foams invariably use cheap fillers to keep costs at reasonable levels and also to make foams heavier. These heavily filled foams may not function as well as less filled foams.

3.6.2 Hardness

Probably the next important physical property of a foam is its hardness or softness. Hardness is measured by its indentation force deflection (IFD). By definition it is the amount of load necessary to press an indenter foot into the foam sample being tested as a percentage of the thickness of the sample, and expressed as a ratio factor. For PUF a factor <2.0 is considered inferior and >2.0 a good foam. The most common physical measurement is called the 'modulus' or support factor. In addition to the 25% IFD, a second IFD measurement is taken with the foam sample compressed to 65% of its original height. The ratio of the two readings – 65% compression reading divided by the 25% compression reading – is the support factor of the foam. The support factor is important in foam cushions for furniture and foam mattresses. This measurement gives an indication of the relationship between its ability to support weight and prevent 'bottoming out'. The higher the IFD, the better is the support. In general, the market does not accept foams with an IFD <1.85.

3.6.3 Foam durability

In general, foam durability applies to flexible foams and is used for comfort purposes. Durability is also thought as 'fatigue softening' (a measurement of the durability of a foam that is extremely important). Density is one of the major controlling factors of fatigue softening.

3.6.4 Compression set

The compression set is the tendency of a foam to remain compressed after the force of compression has been removed. The compression set is another important factor (especially in comfort applications). The magnitude of good values of the compression set and how it is measured depends on different testing methods.

3.6.5 Thermal conductivity

There are two values, the R-value and K-factor, which are important in thermal conductivity. The former determines the resistance of the foam to airflow. The K-factor determines the actual thermal conductivity of the foam.

The R-value is a laboratory-generated value and does not factor-in the other major conditions that can affect the performance of an insulant, such as wind velocity, convection, openings, cracks, and pressure differential between the inside and outside of a building. Both values are important in building insulation to determine the required thickness of the foam material to be used because the thermal conductivity from outside to inside or inside to outside depend on it for effective insulation. As a general rule, low-density foam material needs a greater thickness, whereas high-density foam material can use a lower thickness. In EPS foam boards, this factor is solved by an integral outer skin produced during the extrusion process on both sides.

3.6.6 Water vapour permeability

Water vapour permeability (WVP) is the amount of water a foam material allows to pass through its foam structure. This factor is especially important in outdoor applications or places exposed to high humidity. As the density decreases, the water absorption increases. As a result, open-cell foams are more hydrophilic (water absorbing) than closed foams. Closed-cell foams have a low WVP and, therefore, considered to be hydrophobic (water repelling).

3.6.7 Dimensional stability cold ageing

The more open the cell structure, the better the dimensional stability, whereas higher foam densities also provide higher dimensional stability. The more stretched out or elongated the cell structure the lower is the dimensional stability. If water combines with CO_2 to form foam structures (as in the case of water-blown foams), they produce open cells. Low-density open-cell foams are stable, but closed-cell foams can shrink at low temperatures.

3.6.8 Dimensional stability dry ageing

As ambient temperatures rise, high-density foams tend to expand and swell. Open-cell water-blown foams need to reach equilibrium early in the foaming process. The CO_2 generated during foaming needs to escape the newly formed cell at a rate of equilibrium even to the rate at which atmospheric pressure is pushing down on the foam. Unlike closed-cell foams in which shrinkage can occur later, open-cell foams show shrinkage within the first few minutes. If open-cell foam has a problem with respect to dimensional stability, it can be detected immediately. With closed-cell foams, it may be covered by drywall or wood, and never be detected.

3.6.9 Fire performance

Fire performance is an important factor in building-construction applications. A test helps to evaluate the surface-burning conditions and behaviour of a building material under laboratory conditions. The test measures the flame spread and smoke generated during afire. This is also called a 'fire-retardant factor' and manufacturers of cellular foams incorporate special additives to 'retard' spreading and smoking when these specialty foams come into contact with fire.

3.7 Chemistry of water-blown cellular foams

I present here the chemistry of production of cellular polymer foams with water as the blowing agent. I have selected two of the most versatile and widely used foams for presentation: cellular PU (comfort and automobiles) and EPS (packaging and building construction).

3.7.1 Polyurethanes

The chemistry of production of cellular foams with water as the blowing agent for PUF is very important. It also involves hazardous chemicals and the possibility of fire due to the chemical reactions being exothermic. At the end of this section, Table 3.1 shows the differences in properties if different levels of water are used. Flexible and rigid PUF comprise ≈60% of cellular foams used.

A range of foams from low to high density with good cell structures can be made with water as the sole blowing agent. However, if very-low density and very soft foams are desired, then a small percentage of a secondary blowing agent, such

Table 3.1: Effects of water levels on PUF properties.

Foam idenfication	L3	L5	L7	L9	L11	Description
Water level (phr)	3	5	7	9	11	Number of parts
Cream time (s)	20	27	28	30	35	Chemical reaction
Gel time (s)	49	72	80	85	91	Polymerisation
Tack-free time (s)	62	110	134	200	240	Demoulding
Density (kg/m^3)	46	33	27	24	24	Weight/volume
Average cell diameter (μm)	380	439	498	461	458	Cell size
Cell density (cells/mm)	701	648	558	652	656	Cell weight
Compressive strength (kPa)	330	227	181	164	161	Load factor
Compressive modulus (MPa)	6.3	5	3.9	3.1	3.4	Support factor
Volume shrinkage (%)	0.12	0.18	0.16	0.13	0.11	Temperature
Volume swelling (%)	0.11	0.32	0.23	0.25	0.24	Gases exit

as methylene chloride, can be used. Using water as a blowing agent instead of a chemical helps to reduce environmental concerns considerably. However, when formulating, one should understand the function and limits of the water content in relation to the polyol content and isocyanate content in a formula. In general, it is safe to use ≤6.0 parts of water based on the polyol content in a formulation because higher levels tend to increase the fire hazard during the exothermic chemical reaction between the water and isocyanate. Lower contents of water produce higher densities, whereas increasing the water content produces lower densities.

Laboratory experiments and various studies have shown that the level of water used has important effects on the microstructure of water-blown PUF and, therefore, affects physical–mechanical properties because water initiates the gas for foaming but is also involved with chain extension of a PU matrix. The volumetric change of PUF with low water levels produces excellent dimensional stability. PUF are used for various applications (e.g., insulation, cushioning, bedding, automotive, structural materials, marine equipment, and packaging) due to their excellent physical and mechanical performances and variable molecular designs.

The basic components in a PU formulation consist of a polyol, isocyanate, water, a catalyst, a surfactant and other additives. These components can be foamed immediately by mixing for a few seconds, producing a crosslinking stage, and being allowed to cool. Blowing can be carried out by a PBA (water), a CBA or a mixture of the two. Because of the low thermal conductivity of the gas and small cell structure, PUF prepared with PBA have moderate density and extremely low thermal conductivity that gives rigid foams broad use in insulation and energy-saving applications. Water does not deplete the ozone layer of the Earth and is considered to be the perfect substitute for blowing PUF.

Some researchers consider PUF as composites in which the air bubbles in the foam act as 'reinforcements' for their continuous distribution and that PU resins act as continual polymer matrices. The cellular structure of PUF with air bubbles as reinforcements gives foams with very low density, excellent insulation from heat and sound, and good buffering. Mechanical properties depend on the intrinsic properties of the polymer matrix in call walls and on their architecture as determined by the wall thickness, size distribution and shape of cells. Water blowing of PUF consists of three main chemical reactions.

The reaction between the isocyanate and polyols (polyol system) forms long-chain PU macromolecules through urethane linkages. The reaction of water with the isocyanate produces a carbamic acid that decomposes to yield heat, CO_2 and a highly reactive primary amine. The latter reacts immediately with isocyanate to form a di-substituted urea. The higher the content of water in the formula, the higher is the content of CO_2, heat and polyuria structures generated in the resultant PUF. Heat and CO_2 contribute to the gas–liquid-phase separation in the reactive mixture as well as expansion and solidification of the gas bubbles. Therefore, they have important roles in development of the cellular structure of the foam. Hence, the water content in the formulation has considerable effects on cellular structure and mechanical properties.

PUF with low densities, good mechanical properties and excellent dimensional stability could be produced by fully water-blown technology. When water levels are increased, the cream time, gel time and the tack-free time of foaming mixtures will increase accordingly, while the density and mechanical properties decrease gradually and the cell diameter initially will decrease and the increase. Water levels will have only a slight effect on dimensional stability of PUF because of the strong foam framework of water-blown PUF. Table 3.1 below shows the effects of different water levels on PUF properties.

3.7.2 Water-blown expandable polystyrene

The world's first water-blown expandable polystyrene (WEPS) beads were developed by the new technology centre of Nova Chemicals in Breda (The Netherlands). This revolutionary conventional EPS contains 5–10 wt% of micronised water droplets instead of the standard pentane used as a blowing agent. Perhaps the credit should go to Shell Chemical Company, who initiated research on this subject by funding projects in The Netherlands and Belgium.

The process to make WEPS involves creating a molecular bond between corn-starch and styrene monomer using a compatibiliser such as maleic anhydride. The starch is encapsulated in a shell of PS and it absorbs a small amount of water during aqueous polymerisation. WEPS helps processors to meet the ever-stricter environmental regulations without the capital costs of pollution controls. Another big advantage is that pre-expanded WEPS beads do not need the standard long ageing time of ≥12 h (and often as not 24 h) and can be moulded almost immediately.

Pentane softens PS like a plasticiser, thereby aiding pre-expansion. However, after expansion, the pentane leaves soft bubbles with a slight vacuum inside. These bubbles must refill gradually with ambient air before EPS beads are strong enough to be moulded into desired shapes. Water does not plasticise the PS, so water-blown beads do not soften and can be used immediately as they emerge from the pre-expander: this is a big factor in saving production time. However, WEPS beads cannot be expanded on standard pre-expanders and need slight modification. Pre-expanders for WEPS need a slight vacuum inside that causes the air pressure within the beads to increase to add to the expansion force. Normal pentane-blown EPS beads after being moulded into any shape may need to be stored for some time to allow for shrinkage before being shipped out: WEPS-moulded parts do not.

WEPS requires more energy for pre-expansion than standard EPS. However, in practice, WEPS takes much less energy and time for shape moulding because WEPS requires less cooling in the shape mould after fusion due to the fact that the beads are not softened. If colour in the beads is desired then this can be done during polymerisation.

3.8 Cellular foam tests

A basic understanding of physical properties is important for production of good foamed material (whatever the end applications). Foam manufacturers and their clients should work closely to achieve the desired results by correct formulation and foaming processes. Several international standards are used for testing for a wide range of parameters important to foam manufacturers and end-users, one of which is the American Society for Testing Materials (ASTM). Some of the important tests are:

- Density (ASTM D-1622)
- Closed-cell content (ASTM D-2856)
- Open-cell content (ASTM D-6226)
- Compressive strength (ASTM D-1621)
- Tensile strength (ASTM D-1623)
- Thermal conductivity (ASTM C-518)
- WVP (ASTM E-96)
- Dimensional stability cold ageing (ASTM D-2126)
- Dimensional stability dry ageing (ASTM D-2126)
- Dimensional stability humid ageing (ASTM D-2126)

Definitions are:
- Density of foam is defined as the weight per unit volume.
- Closed-cell content is measurement of the number of intact cell membranes (windows) that prevent no open passageway for air to flow. For rigid closed-cell foams, >90% of the cell windows are closed.
- Open-cell content is measurement of the number of broken cell membranes (windows) that allow air passageway for airflow. For open-cell foams, >90% of the cell windows are open.
- Tensile strength is a measure of the amount of stress required to break the foam as it is pulled apart.
- Compressive strength is a test used to determine the value of maximum compressive force a foam material can withstand.
- Thermal conductivity indicates the amount of airflow (heat or cold air) that the insulant allows whether it is from outside to inside or from inside to outside.
- WVP is a laboratory test used to determine the rate of infiltration of water vapour through insulating materials.
- Dimensional stability cold ageing is a measure of change in dimensions upon exposure to extreme conditions.
- Dimensional stability dry ageing is a test to determine possible expansion and swelling due to increase in ambient temperature.

Bibliography

1. T. Chen, Y. Mi, H. Du DU and Z. Gao in *Mechanical Properties and Dimensional Stability of Water-Blown PU Foams with Various Water Levels*, College of Material Science and Engineering, Northeast Forestry University, Harbin, China.
2. *Foam Blowing Agents*, United States Environmental Protection Agency, Washington, DC, USA, 28th October 2014.
3. A.M. Helmenstine in *Water Chemistry Facts and* Properties, About Inc., 2016. http://chemistry.about.com/od/waterchemistry/a/water-chemistry.htm.
4. *Waterborne Polyurethane Market by Application & Geography*, MarketsandMarkets, Magarpatta City, India, 15th May 2015.
5. *EPS Processing Machinery, Packaging & Equipments*, Mane Electricals, Pune, India, 2011 [Private Communication].
6. J.H. Schut in *Water-Blown EPS Will Help You and the Environment*, Plastics Technology, Cincinnati, OH, USA, 2002.
7. C, Defonseka in *Practical Guide to Flexible Polyurethane Foams*, Smithers Rapra, Shawbury, UK, 2013, p.150.
8. http://scifun.chem.wisc.edu/chemweek/polymers/polymers.html.

4 Processing methods for water-blown cellular polymers

4.1 Expandable polystyrene

Conventional expandable polystyrene (EPS) in the form of beads can be incorporated with tiny amounts of petroleum-based gases as blowing agents during polymerisation. Under heat and pressure, each bead can be expanded to form cellular structures. Over the years, these blowing agents have been banned from use due to their depleting action on the Earth's ozone layer. Subsequently, pentane was used as the blowing agent, which was less harmful to the environment and continues today. Bead size can vary from 0.1 mm (small) to 0.3 mm (larger beads) and is determined at the time of polymerisation. Most grades are not coloured and give a 'white' effect on expansion and moulding, but some manufacturers also produce coloured grades. A wide range of grades are available to suit different end-applications such as building insulation, domestic/consumer products, acoustic/sound absorption, and marine applications. To produce these applications, in addition to a blowing agent and colouring, other additives are incorporated at the time of polymerisation.

Due to research and development as well as to ease concerns regarding the environment, scientists have created a unique 'blowing' system of using water as the sole blowing agent, even eliminating the use of pentane. Surprisingly, this is providing many additional advantages over conventional blown polystyrene (PS) with pentane, which has been mentioned earlier in this book. These new-technology EPS beads are called 'water-blown expandable polystyrene' (WEPS). The standard packaging for these beads has not changed and they are available in packs of 25 kg (paper bags), 250 kg (steel drums) or in bulk packs of ≈400 kg (gaylord). This new product has a longer shelf-life unlike the gas-blown beads, whereby the containers must be kept sealed/closed at all times to prevent the gas from escaping upon atmospheric exposure (which can also create a fire hazard on factory floors in the long-term due to static electricity).

4.1.1 Processing methods

Selection of the size and grade of beads is dependent upon the end-product. For example, if the end-product is to be the manufacture of moulded boxes (e.g., fish boxes), beads of smaller size should be selected, along with medium-to-larger sizes for large solid blocks to be cut into sheets. For fishing floats and floating devices, the smallest size of beads provides strength. It is possible to mix two or three sizes of beads to achieve strength, light weight, or cost reduction. Selection of a suitable

https://doi.org/10.1515/9783110643121-004

bead size for processing for a particular product or end-application depends greatly on the final density of the product to be achieved.

Standard pentane-blown beads must be stored at ≈21.1 °C (semi-cool room) to prevent the gas in the beads from diffusing slowly at room temperature. Even with these precautions, the expected shelf-life is ≈6 months. This is not the case for WEPS, but it is prudent to store beads in a cool place away from excessively warm areas. The first stage of operations is to pre-expand the beads using a supply of unsaturated steam (saturated steam burns/scorches beads). The range of full expansion of beads is ≈40–50% of their original size, and the first stage expands them to about half of their original size. Standard pre-expanders cannot handle WEPS beads and a special type of pre-expander is needed otherwise the existing beads must be modified. Then, the semi-blown beads go through an agitator bed/fluidised bed to remove lumps that may a have formed before being stored in large vertical silos.

Unlike pentane-blown beads, which must be allowed to cool and mature (stabilising of cell walls) for ≥24 h, these WEPS pre-expanded beads can be moulded immediately.

Moulding can be in the form of manual, semi-automatic or fully automatic operations in which the pre-expanded material is fed into aluminium moulds in two halves. In the case of semi-automatic or fully automatic moulding, one-half of the mould is fixed on one platen and does not move. The other half is closed (in general it is mounted horizontally) by pneumatic, hydraulic or electric movement and the material is filled through an opening. Use of filler-guns ensures that the mould is filled fully and excess material reverses back into the feed source.

For production of large blocks, moulds (which are called 'block moulds') with a lid opening in which the material is filled and then closed tightly are used. In both operations, steam is sent into the material through a uniform set of inlets to reach all the material uniformly and simultaneously. Then, the material is expanded fully and takes the shape of the mould. In multiple-cavity moulding, fill-guns fill the mould simultaneously and uniformly. During the short steaming process, the pre-expanded beads are expanded further until they fuse together under heat and pressure. External and internal gauges control this phase and cut-off to pre-set parameters automatically, followed by a cooling cycle. If water is used as the cooling medium it takes longer to cool but, if vacuum cooling is used, the cycle is much shorter. Opening of the mould can be done manually or on automatic mode, and the moulded product is ejected. The operator must ensure sufficient cooling of the moulded product because insufficient cooling and opening of the mould prematurely results in distortion or shrinkage of the product. The hot cells must be stabilised sufficiently before being exposed to the atmosphere.

If cooling is with water, then moulded products should be stored for ≈20 h before shipping.

For production of sheets for insulation, large blocks are made from which sheets of different thicknesses can be cut using hotwire systems. Cutting can be done by a single wire or multiple hotwires whereby and entire block can be cut in one pass. Some blocks may need trimming of all sides, some sides, or 'skins' before cutting sheets. Advanced block-moulding machines are available that can give perfect cut surfaces without the need for trimming. Production of EPS sheets with skins and extruded PS for insulation of external buildings is not discussed here because they are not water-blown.

Moulding machines generate hardly any waste, but waste can result from incorrect parameters being set, trials being conducted, or due to power failures. In this case, waste can be granulated and mixed with pre-expanded material ≤10–15% and brought into the main stream.

4.1.2 Machinery and equipment

Selection of machinery and equipment needed to process WEPS polymers depends entirely on the mode of operation. The correct type of pre-expander and a suitable steam source (boiler) to provide a good steady supply of quality steam are crucial. The following information provides sound knowledge of what is required to set up a medium-size plant:

- Products: WEPS blocks, fish boxes, fishing floats.
- Estimated floor space: 1,000–1,200 m^2.
- Steam boiler: capacity 600 kg/h (oil fired) 100 psi (pressure) with water softener. Insulate the steam pipes to prevent excessive condensation on steam lines. However, the steam supply lines leading to and from the main lines to the moulding machines should not be insulated to prevent a supply of superheated steam to the moulds.
- Fully automatic pre-expanders for WEPS need to withstand higher temperatures and pressures than those for standard pre-expanders, so thermocouples and pressure gauges must be added. WEPS is pre-expanded at 120–130 *versus* 100–115 °C for conventional pentane-blown EPS. Pre-expanders with microwave energy heat the water droplets within the beads, not the plastic itself. Then, heat transfers from the boiling water to the PS, which expands as it reaches its glass transition temperature (T_g). Application of steam in a pre-expander heats the polymer first and then transfers from the polymers to the water but this takes longer and water vapour is lost. Overheated foam also takes longer to cool, thereby allowing bubbles to collapse partially during cooling.
- Size of a fully automatic hydraulic shape-moulding machine is 1.4 × 1.2 × 1.2 m.
- Size of a semi-automatic block-moulding machine is 2 × 1 × 0.5 m (thickness).
- Auto pneumatic material-filling gun with collet.
- Four-cavity aluminium mould for fishing floats.
- Two-cavity aluminium mould for fish boxes plus lid.
- Galvanised iron-fabricated Nylon-mesh silos (10 m^3 capacity).

- 5-HP water pump.
- Vacuum-cooling system or water-cooling system and air compressors.
- Steam lines and other parts.
- Post-moulding cooling: if vacuum cooling is not used but water cooling is used, then a re-circulatory cooling system for warm/hot water can be used to prevent waste of water. However, these waters may have to be replaced periodically if they become dirty or discoloured.
- Essential spare parts.
- Three-phase electrical power.
- Cutting machines: If WEPS blocks are made, then at least two cutting machines are required: a vertical cutter and horizontal cutter. Standard simple hotwire cutting machines with a power controller (0–100 V) for nickel chrome thin gauge wire is sufficient. More sophisticated cutting systems (e.g., laser cutting, band-saw cutters or water-jet cutting) can also be used but most producers use hotwire systems.
- A granulator is optional but helps to re-use ≈15% of waste for making blocks.

The information given above is the basic required but can be varied to suit product range and or increase production volumes by increasing the number of machines in relation to available steam supplies. A typical floor layout plan for installing machinery to ensure smooth production flow is shown in Figure 4.1.

4.1.3 Troubleshooting

As in any manufacturing process, there are bound to be moulding faults caused by unsuitable/defective equipment or incorrect parameters in the processing itself. Remedial actions that a processor of WEPS can take to ensure good products are given in Table 4.1.

4.2 Flexible polyurethane foams

Here I discuss the principles, formulation, processing methods as well as the basic machinery and equipment required to make flexible polyurethane foam (PUF) with water as the sole blowing agent. The large-volume, more sophisticated processing methods and equipment are not presented because such formulations contain 'blowing systems' rather than water as the single blowing agent.

Basic raw materials for making flexible PUF are polyols, isocyanates, water, amine catalysts, tin catalysts, chain extenders, surfactants, colourants and additives (e.g., fire retardants, antioxidants, and ultraviolet (UV) protectors). Three of the most important properties of a foam are the final density, cell structure and

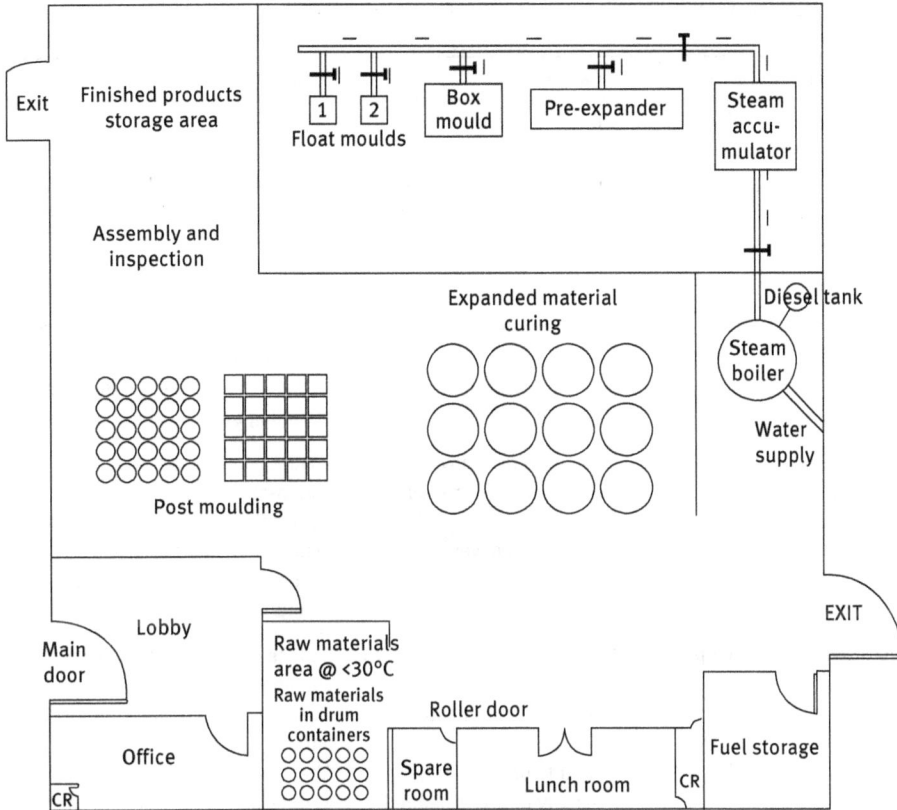

Figure 4.1: Layout of an EPS plant. Project designed and setup by the author in the Philippines and drawn by Melvin L. Motas. Note: 1–2 metric tonnes/month and floor space: 5,000 ft² (not to scale).

support factor. For a producer of foams, understanding the function of each component is crucial to formulate correctly and achieve the desired end results, while keeping profitability in mind. Processing equipment and the parameters used have a considerable effect in achieving good foams. It is good practice to produce foams conforming to a chosen international standard for ease of marketing.

4.2.1 Polyols

Most flexible PUF are made with polyether polyols, which are based on petroleum. Over the last few years, polyols from various other natural sources (e.g., vegetable oils) have been appearing on the market and are being used by some manufacturers for the production of flexible PUF. However, the yields from these new polyols are less than those obtained from conventional polyols. The common petroleum-based

Table 4.1: Remedial actions for mould processing faults.

Moulding fault	Cause	Remedy
Poor welding (poor adhesion of beads)	Density of pre-expanded beads is too low	Increase density/increase fill
	Pre-expanded beads not of uniform density	Check pre-expander/expansion parameters
	Insufficient steam or steam at too low a pressure	Increase the cross-section of steam line/increase pressure/increase inlet slots
	Condensate collects in the mould	Increase pre-heating time/check condensate valves
	Too rapid welding of the outside of the product	Reduce steam pressure/increase steaming time
Excessive moisture content in product	Inadequate removal of condensate from mould	Ensure mould is pre-heated /drain-off condensate appropriately before moulding
	Steam is too wet	Insulate steam line; install a condensate separator before moulding
Distorted or shrunken products	Moulding steam pressure is too high	Reduce pressure in the steam chamber
	Steam is superheated	Take measures to see steam is dry and not overheated
	Mould not filled fully	Check and increase fill
	Density of pre-expanded beads is too low	Reduce pre-expansion and check density required
Blocks are to heavy	Pre-expanded beads are too heavy	Check and reduce the density
	Mould is overfilled	Reduce filling
	Excessive moisture content	Too much retention of water in mould
Blocks are too light	Pre-expanded beads are not dense enough	Increase the density of beads
	Mould not filled completely	Increase fill and ensure that the mould is filled completely
Contoured mouldings — Inadequate welding/adhesion	Mould walls are too cold	Pre-heat the mould sufficiently. Activate condensate valves and remove water
	Inadequate distribution of steam	Provide extra steam inlets for better distribution

(continued)

Table 4.1 (continued)

Moulding fault	Cause	Remedy
Localised shrinkage	Mould is not filled appropriately	Ensure bead sizes can enter the narrower parts of the mould readily
	Mould is too wet	Purge and pre-heat
	Pre-expanded beads are not dense enough	Use denser beads (particularly if fast cooling cycles are used)
Moulding bulge	Insufficient cooling of the mould	Increase the cooling period. Check and use cooler water

polyols are made by copolymerising a mixture of propylene oxide and ethylene oxide with glycerine as the initiator.

Different types of polyols with different functionalities give different properties. Polyols are liquids and are available commercially in drums, totes or tankers depending on the size of the operation. Their shelf-life is ≈6 months and, for optimum yield, should be kept at ≤25 °C. Precautions should be taken to prevent water absorption, especially from the atmosphere. All formulations are based on the content of polyol in the formula, taken to be equivalent to 100. In general, producers know which grade to use but can also be guided by the suppliers or manufacturers to suit the end-application.

4.2.2 Graft polyols

Graft polyols contain copolymerised styrene and acrylonitriles. They are designed to maximise loading properties in polyurethane (PU) formulations, contain solids in the range 10–45%, and may also be considered 'filler' polyols.

4.2.3 Bio-polyols

Global environmental concerns have led to the need of alternate non-petroleum-based sources for polyols, and research has led to the creation of bio-polyols. The latter are polyols derived from vegetable oils such as soya, canola, and peanut oil, and are made by different methods. These polyols can be used as primary polyols in a formulation but the foam yields may be lower. All are clear liquids, ranging from colourless to light-yellow, with varying viscosities that are a function of their molecular weight (MW). Odour varies from polyol to polyol, but they carry a faint odour of the original source of oil and tend to turn rancid during storage. During foaming, use of an antioxidant neutralises the odour in the final foam.

4.2.4 Isocyanates

The most common isocyanate used in flexible PUF is toluene diisocyanate (TDI), though diphenyl methane diisocyanate and its polymeric forms can also be used. Mixtures of 2,4-TDI and 2,6-TDI are the isocyanates of choice for flexible foams. TDI is available in isomer mixtures of 80:20 or 65:35 as well as a pure 2,4-TDI isomer, but the reactivity of the systems and resulting foam properties can be modified using blends of the various isomers. TDI is a low-cost, high-quality product that allows foam manufacturers to produce many types of flexible foams with a wide range of physical properties.

4.2.5 Blowing agents

Water is sufficient as a sole blowing agent to produce good foams of densities $\approx$18–32 kg/m^3. However, for very low densities, an auxiliary blowing agent such as methylene chloride can be used. Other blowing agents can also be used but some are banned due to environmental concerns. Water is present in every PUF formulation but has a maximum threshold limit due to excess generation of heat during the exothermic reaction of components during foaming. Water reacts with isocyanate and forms carbon dioxide (CO_2), which also acts as a blowing agent. As a general rule, the maximum safe water component in a PU formulation is $\approx$6.0 parts by weight of polyol. Lower contents of water produce higher foam densities, and higher contents yield lower densities.

4.2.6 Catalysts

Production of flexible PUF requires use of catalysts because two major reactions take place. In the polymerisation or gelling reaction, poly-functional isocyanate reacts with polyol to form PUF. In the gas-producing or blowing reaction, the isocyanate reacts with water to form polyuria and CO_2. These reactions occur at different rates, with both reactions dependent upon temperature, catalyst levels, catalyst type and various other factors. However, to produce high-quality foams, the rates of both reactions must be controlled and balanced.

If the gas-producing reaction (blowing) occurs faster than the polymerisation reaction (gelling), the gas generated by the reactions may expand before the polymer is strong enough to contain it, so internal splits and foam collapse can occur. In contrast, if the polymerisation reaction occurs faster than the gas-producing reaction, the foam cells remain closed, causing the foam to shrink as it cools. Tin catalysts (usually stannous octoate) are used for PUF production for the polymerisation reaction, whereas tertiary amine catalysts are used for the gas-producing reaction. If these reactions are well-controlled and balanced, good foam structures with uniform cell structures can be produced.

4.2.7 Surfactants

Non-ionic silicone-based surfactants are used in the production of flexible PUF. Different grades of surfactants are available to foam manufacturers to meet specific needs. Main functions of surfactants are:
- Reduction of surface tension.
- Resilience to prevent foam collapse when foam is rising.
- Surfactants control cell size for uniformity.
- Silicones counteract the deforming effect of solids in the reacting system.
- Surfactants prevent bubble breakage at full rise of foam.

The most important factor is stabilisation of cell walls. Surfactants prevent the coalescence of rapidly growing cells until they have attained sufficient strength through polymerisation to become self-supporting.

4.2.8 Fillers

Ideal fillers for PUF are inert inorganic compounds of very fine particle size. They are added to foam formulations to increase density, load bearing, and sound absorption as well as to help cost reduction for foams. Use of filers may affect certain physical properties of a foam. Of the wide range of fillers available, calcium carbonate is the most popular and widely used filler.

4.2.9 Pigments

Foams made from general foam formulations are colourless (white). A yellow colour is used to protect the foam from UV action (which discolours the foams with slight degradation). Foam manufacturers use different pigments to colour foams for easy identification of density, and coloured foams helps the aesthetics of products. Polyol and water-based pigments or compatible pigments are preferred. If water-based pigments are used, care must be taken to make necessary adjustments in the pure-water content in the formulation so as not to go over the threshold limit for water. Some typical problems that maybe encountered using pigments are: foam instability, foam scorch, colour migration as well as abrasive action on pumps and mixers.

4.2.10 Additives

Additives are materials that are added to foam formulations to achieve the desired end-properties. They do not interfere or affect the general chemistry. Some common additives are:
- Fire retardants
- Antioxidants
- Cell openers
- Plasticisers
- Antibacterial agents
- Anti-static agents
- UV stabilisers
- Foam hardeners
- Crosslinkers
- Compatibilisers

4.2.11 Typical formulations for water-blown flexible polyurethane foams

In PUF production, the different components can be formulated with different quantities to achieve foams with different properties to suit end-applications. Table 4.2 shows the basic formulations used for making foams for cushions, mattresses and pillows.

Table 4.2: Typical formulations for mattresses and pillows.

Component	Standard	Mattress	Pillows
Polyol	85.00 pbw	80.00 pbw	100.00 pbw
Graft polyol	15.00	20.00	–
TDI	40.70	70.00	84.00
Water	2.90	3.50	4.50
Surfactant	1.20	0.60	0.30
Amine catalyst	0.12	0.70	0.27
Tin catalyst	0.04	–	–
Methylene chloride	–	–	–
Flame retardant	3.00	–	–
Density (kg/m3)	27.20	20.80	17.60
Support factor (IFD)	2.30	2.10	2.00

All formulations for cushions and pillows by the author
IFD: Indentation force deflection

4.2.12 Processing technology

Properties of all water-blown conventional polyether PUF are controlled more by formulations and processing conditions than the properties of the individual components. Therefore, only the major effects of water, isocyanate (index) and polymer solids need to be considered. The term 'polymer solids' is used interchangeably with 'parts of polymer polyol' even though these two items differ by a constant factor. Processing conditions include those that affect foam breathability, cell structure and cell count. The most important properties of PUF are the density and support factor (IFD).

The general rule is that, as water content increases, the density and IFD decreases. Higher contents of isocyanate in a formula give softer foams and lower densities and *vice versa*. This phenomenon is dependent on the amount of polymer polyol used and the actual amount of water used as the blowing agent. At densities corresponding to 2.5–4.0 php water, IFD increases with increasing water. If >30 parts of high solid polymer polyols are used, foam firmness decreases with increasing water content.

Processing of components to make foams can take the form of manual, semi-automatic or continuous foaming operations. To avoid unnecessary waste, a small volume in-house box test be done to test the suitability of the formulation before production runs. There can be slight variations of processing sequences. However, as an example, let us use a semi-automatic operation in which all components in different vessels are connected to a central mixing head with a mixing vessel that lies on the floor of a mould of pre-determined size (e.g., round, square or rectangular). First, the polyol is brought on-stream and mixed with colour for a few seconds. Then, the other components other than the isocyanate are brought on-stream, introduced into the mixing vessel, and mixed with the polyol. Finally the isocyanate is brought direct into the mixing vessel and mixed. This must be done between 4 and 6 s and the full mixture released onto the floor of the mould by auto-opening of the bottom of the mixing vessel. If this time limit is passed the mixture starts polymerising and foaming inside the mixing vessel, and the whole batch is wasted.

The mixture, still in a liquid state, spreads evenly across the floor of the mould and starts 'creaming' (polymerisation), and the foam rises slowly. If formulated appropriately, the foam rises gradually to a few inches of the top of the mould and, within a short period, cure and form a 'skin' on top which is tacky. The foam block is allowed to settle for a few minutes and, when the foam is no longer tacky, it can be de-moulded. If the formulation is faulty, defects such as splits, surface bubbles, and foam-block collapse occur. Then, these blocks are stored in a cool place and positioned ≥1 foot apart because heat is generated in the final curing time due to the exothermic reaction taking place. Freshly foamed blocks must never be stacked on top of each other because a fire can be started. After ≈24 h, these blocks can be taken for fabrication. For a full detailed account of the technology of flexible foam, see [1].

4.2.13 Processing machinery and equipment

Water-blown flexible PUF can be processed on any conventional foaming systems because production of good-quality foams depends mostly on correct formulation to ensure correct chemical reactions and achieve the desired end-properties.

4.3 Water-blown high-resilience foams

Open-cell PUF, though belonging to the same PUF family, have random cell structures unlike the uniform cell structures of flexible foams. These foams are denser, have better resilience and support factors, and recover faster than even viscoelastic foams when compressed. They are high-end foams and used widely for added comfort, especially in furniture, bedding and specialty applications. The chemicals used for making them are also more expensive than those used for standard foams.

Here, I describe an interesting water-blown two-component system. Some manufacturers of raw materials offer two-component systems for ease of processing. However, a foam producer will be limited to making foam of one density and properties confined to that particular system. Consider the water-blown two-component system HR 250 manufactured by Greenlink Incorporated (Melbourne, Australia). HR 250 can be mixed using an electric drill or similar device with variable speed. At 3,000 rpm, a good homogeneous mixture can be achieved and is ideal for an operation of small-to-medium size to suit an entrepreneur with limited experience or knowledge and a small operating budget. Table 4.3 shows the important production parameters for this particular two-component HR250 grade.

Table 4.3: Production parameters for a two-component system.

Mixing speed	3,000 rpm
Mixing ratio	Polyol:TDI = 100:57
Mixing time (s)	7–8 s
Cream time (s)	12
Gel time (s)	70
Tack free time (s)	300
Free rise density (kg/m^3)	57

By increasing or lowering the isocynate content, a foam producer can make harder or softer foams, respectively. If colour is used, add a desired amount to the polyol blend and mix well before incorporation of the isocynate. Recommendations based on the author's production experience.

4.4 Water-blown rigid polyurethane foams

Density is an important parameter that influences the properties and performance of water-blown rigid PUF. If variations of water are used as the blowing agent in the production of rigid foams resulting in different densities, these differences will affect the mechanical, morphological, water-absorption, thermal-conductivity and thermal behaviour because densities control foam architecture. Laboratory tests have shown that the density of rigid foams decreases from 116 to 42 kg/m^3 with an increase in water content from 0.1 parts per 100 polyol by weight to 3.0 parts per 100 polyol by weight, respectively. Water absorption increases with decreases in density due to the increase in cell size and decrease in cell-wall thickness. Thermal conductivity decreases with increases in density due to decreases in cell size. The T_g increases with decreases in foam density, but thermal stability decreases with decreases in foam density.

Rigid PUF are very popular energy-efficient and versatile insulants. These foams can cut energy costs significantly while making commercial and residential buildings more efficient and comfortable. In a typical home in the western world, ≈60% of energy used is for heating and cooling, making it the largest energy expense for most homes. To maintain uniform temperature and lower noise levels in homes and buildings, builders use rigid PUF. These foams can be used as insulation for roofs, walls, windows, doors and as air barriers.

4.5 Water-blown integral skin polyurethane foams

Integral skin foams (ISF) consist of a cellular core and a solid outer skin. ISF are produced in one operation of foaming. Ban of the use of chlorofluorocarbons as a blowing agent ensures that ISF could not be made with water as the blowing agent because the CO_2 generated by an isocyanate–water reaction is unlikely to condensate at the foam surface. However, new technologies have emerged to make all-water ISF. An example is a novel polypropylene glycol that contains an extremely small amount of byproducts having a narrow MW distribution when applied to conventional systems. This polypropylene glycol balances between the gelling reaction and blowing reaction and produces a tight outer surface of the foam that is as good as the skins formed when banned blowing agents were used.

ISF cover a wide area of applications in the automotive industry. In addition to the foams that make car seats comfortable, bumpers, interior 'headline' ceiling sections, the car body, spoilers, doors and windows use PUF. These foams also give drivers and passengers more automobile 'mileage' by reducing weight and increasing fuel economy, comfort, corrosion, insulation and sound absorption. The formulation can be adjusted to produce skins with different thicknesses to suit end-applications, and water can be adjusted to produce different densities.

4.6 Open-cell *versus* closed-cell polyurethane foams

Open-cell foams are soft and used mainly for comfort applications such as bedding, cushions, and pillows. Cell walls or the surface of foam bubbles are broken (open) and air fills all the material, thereby making the foam soft and providing a cushioning effect. The insulation value of open-cell foam is related to the insulation value of the calm air within the matrix of broken cells or open cells. The densities of open-cell foams can be 16–32 kg/m^3. Due to the air permeability of open-cell foams, the insulation value is less than that of closed-cell foams.

Closed-cell foams have varying degrees of hardness that are dependent upon density. A normal closed-cell insulation or flotation PU is 32–48 kg/m^3. It is strong enough for an average person to walk on without major distortion. Most of the bubbles or cells in the foam are not broken. They represent a closely clustered bunch of inflated soccer balls, piled together in a compact configuration. This setup makes it strong or rigid because the bubbles are strong enough to take a lot of pressure. The cells are full of air but cannot permeate due to the closed walls.

The advantages of closed-cell foams compared with open-cell foams include their strength and higher R-value (though the cost is higher than that of open-cell foams). The choice of foam and grades depend on the end-applications. The two main areas are comfort and insulation, with closed-cell foams costing more.

4.7 Troubleshooting: Some common defects/solutions in foams

Manufacture of flexible PUF is based on chemical reactions, so a foam producer will experience some defects at some point. There is an inherent waste factor at ≈10–12% due to the outer skins of large blocks having to be trimmed, but these can be converted into other products such as carpet underlay and compressed sheets, which will provide a 'kickback' against profit losses. Defects can occur due to: inaccurate weighing/metering; incorrect mixing times; inferior or contamination of raw material components; faulty machinery; and equipment; power failures; atmospheric conditions.

Some of the common defects in flexible PUF manufacture are shown in Table 4.4.

Table 4.4: Common defects in manufacture of flexible PUF.

Defect	Description	Recommendation
Bottom cavitation	Bottom eaten away	Look for errors in metering
		Decrease tin catalyst
Dense bottom skin	Thick bottom layer of foam	Increase silicone level
Smoking	Excessive TDI vapours	Reduce isocyanate level
Tacky bun surface	Surface sticky for too long	Look for errors in metering
		Increase catalyst level
Flashing/sparklers	Risk of effervescing foam	Decrease isocyanate level
		Decrease silicone/amine levels
		Increase tin catalyst
		Reduce processing temperatures
Gross splits	Vertical/horizontal splits	Increase tin catalyst
		Decrease amine/water levels
		Increase silicone level
Moon craters	Small pockmarks on bun	Reduce air entrapment in pour
		Minimise splashing on pour
Pee holes	Small spherical holes	Increase silicone level
		Reduce mixing speed
		Reduce tank agitation
		Reduce air entrapment
Relaxation	Foam rises and goes down	Increase tin catalyst
		Increase silicone level
		Reduce level of amine catalyst
		Reduce mixing speed/nucleation

4.8 Safety factors for cellular foams

PUF manufacturing calls for in-house safety factors because this process deals with highly corrosive, inflammable and dangerous-to-handle chemicals. Safety precautions must be established and observed at all times, especially by the factory personnel who come into direct contact with these chemicals. Factory safety systems start with water sprinklers, fire alarms and fire extinguishers installed at strategic places. A spill-management procedure is recommended and periodic training of personnel by professionals (e.g., fire brigade, and companies specialising in management of chemical spills) ensure good performance by factory personnel on correct procedure in case of a fire and correct handling of equipment.

Most large-volume manufacturers of foams have in-house eye-wash stations, emergency showers and constant training in safety handling procedures, along with safety posters inside the factory reminding all personnel of a 'safety first' policy. Air quality is important and foam producers should have periodic tests done by professional to ensure air contamination is within established safety limits. Some of the general causes for safety problems are: static creating sparks chemical spills; electrical short circuits; faulty chemical formulations; bad air quality due to inadequate exhaust systems; ineffective maintenance; chemical leaks; personnel not wearing appropriate protective clothing (common occurrence).

Bibliography

1.   J.H. Schut in *Water-Blown EPS*, Plastics Technology, Cincinnati, OH, USA, September 2002.
2.   *Processing Faults for EPS (Styropor)*, BASF Technical Bulletin, Ludwigshafen, Germany.
3.   *EPS (Thermocole) Processing Machinery*, Mane Electricals, Pune, India.
4.   *Natural Choice Polymer Design*, Natural Polymers LLC, West Chicago, IL, USA. http://www.Naturalpolymersllc.com/.

Reference

1.   C. Defonseka in *Practical Guide to Flexible Polyurethane Foams*, Smithers Rapra, Shawbury, Shropshire, UK, 2013.

5 Water-blown specialty cellular foams

Ever since scientists pointed out the alarming trend of global air pollution due to harmful gases being released into the atmosphere, all countries have shown concern about finding solutions. Apart from the industrial and automobile emissions which are formidable, the manufacture and processing of plastic polymers also contributes a major share of air pollution. Due to this factor and to combat rising costs, industrialists and manufacturers have been involved in intense research to find solutions such as: methods for weight reduction; savings in raw materials; moving away from use of petroleum-based sources. Two of the most effective solutions for some of these areas are: i) cellular materials and polymers from naturally occurring vegetable sources (e.g., vegetable oils); and ii) water-blown cellular polymers (where water acts as a blowing agent instead of the conventional petroleum-based ones that are harmful to the atmosphere).

Examples of water-blown specialty foams are: viscoelastic foams; spray polyurethane foams (SPF); hydrophilic and hydrophobic foams; polyisocyanurate foams; soy–phosphate rigid foams; integral skin polyurethane foams (PUF); convoluted and contoured foams; hybrid urethane foams; reticulated PUF; and sulfone cellular foams.

5.1 Viscoelastic foams

Viscoelastic ('memory') foams have special properties unlike conventional two-dimensional (2D) foams. They have four-dimensional properties (density, temperature, hardness and time), all of which can be varied to suit different end-applications. The general density range for viscoelastic foams is 48–96 kg/m^3. A foam producer would need experience and in-depth knowledge for formulation because viscoelastic technology is complex and special polyols (non-conventional) must be used. Selection of the grade of polyol depends on the final desired density, and additives decide the desired properties to suit end-applications.

A special feature of viscoelastic foam is that it is 'lazy' or slow to recover with high damping. One of the main characteristics is its very low resilience, meaning that there is no 'spring' in the foam. Another important aspect is the temperature-sensitivity of the foam. If one were to 'sink' a hand into a surface of viscoelastic foam, it leaves an indentation, which disappears as the foam recovers slowly.

Unlike conventional flexible foams which are 2D (density and hardness), viscoelastic foams have two additional dimensions, all of which can be varied to achieve different properties to suit end-applications. Most flexible foams are highly resilient, 'bouncy' and cannot retain a shape, whereas viscoelastic foams allow moulding to a body shape. The low indentation force deflection (IFD) of these foams allows a body to sink in deep into the foam, cushioning the body without backward resistance (pressure) and thus providing maximum comfort. Viscoelastic foams are heavier than

https://doi.org/10.1515/9783110643121-005

conventional foams and, during production, whatever processing method is used (continuous foaming or production of single blocks), the ideal height should be ≤40 inches in consideration of post-cure handling and transport to the curing area. Unlike conventional flexible PUF in which the post-curing period is ≈24 h, these foams need ≥48 h. Standard safety precautions must be followed because during this period heat will be given out due to the exothermic chemical reactions taking place within the foam.

Some suppliers of raw materials offer a two-component system, component A and component B. A two-component system is easy to process and inexperienced foam producers and entrepreneurs should use these systems (at least at the start). A two-component system is easy to measure and process, but has a distinct disadvantage. Only one particular density (as supplied by the raw-material supplier) can be made whereas, if raw materials were purchased separately, a foam producer could vary formulations to achieve different properties.

Table 5.1 shows typical processing parameters for a standard two-component system. Such a system can be processed on machines that dispense high and low pressure.

Table 5.1: Mixing parameters for a water-blown two-component system.

Density	Free rise	40–56 kg/m^3	2.5–3.5 lbs/ft^3
Handling properties			
Mixing ratio	By weight	Part A	50 pbw
		Part B	100 pbw
Viscosity	At 25 °C	Part A	205 cps
		Part B	550 cps
Cream time	At 25 °C	–	25 s
Rise time	At 25 °C	–	2.5–3.5 min
Demould time	At 25 °C	–	15–30 min

Reference from Barnes Guide 2010 (free download)

5.1.1 End-applications

Viscoelastic foams have revolutionised the market for bedding and comfort products and have all but replaced bedding applications due to their sheer comfort. Moreover, due to their therapeutic value (especially as a stress-relieving surface), several different combinations with other foams are available. Some common end-applications are mattresses, mattress toppers, sheets, wedges, pillows, footwear, wheelchair cushions, medical applications, hospital mattresses, and space travel (to counter G-forces).

5.2 Spray polyurethane foams

Many end-users are confused as to which type SPF insulation should be used. A common difference of opinion occurs when deciding the right type to use: open cell or closed cell. Not all SPF insulations are the same. In general, open-cell foams have a density of 8.0 kg/m^3 and closed-cell foams have a density of 16.0 kg/m^3. The choice may depend on the climate and geographic area.

Spray processes provide control over thickness and flexibility in the application of PUF and coatings to horizontal and vertical surfaces. This is particularly beneficial in applications where large areas must be coated for insulation (e.g., buildings or storage tanks). Conventional spraying equipment is economical, and portable spraying units provide an easy way to insulate and seal structures. This is a key factor in controlling costs. If necessary application of an approved urethane coating is sprayed on top of urethane foam to provide long-term weather protection from elements such as moisture and ultraviolet (UV) rays.

Another advantage of spray urethane foams is the capability to apply spray coatings to oddly shaped surfaces. In general, a two-component system is used by pumping through a spray gun. The pre-mixed components are auto-mixed by atomising the components as they exit the mixhead of the gun at high velocity as very fine droplets. The adhesion strength of PUF is affected by the cleanliness of the surfaces. Moisture, loose gravel, grease and other foreign matter adversely impact bonding of the foam to the surfaces being sprayed. These spray foams are available at high and low densities and can also be custom-formulated to suit the specific needs of spray application (especially with high fire-retardant properties).

Some typical applications of spray foams are:
- Roofing insulation
- Sliding insulation
- Floors
- Wall cavities
- Spas
- Stage scenery
- Displays and attractions
- In-mould applications

5.3 Hydrophilic and hydrophobic polyurethane foams

Hydrophilic and hydrophobic PUF are specially formulated PUF for industrial work (mainly as sealants). These foams, if used in sealing or repair of concrete, are said to be 'hydrophobic' (water repelling) or 'hydrophilic' (water absorbing). Both types can be used as sealants but hydrophobic systems are best suited for permanent repair for most leaking applications (unless there are structural issues).

A hydrophobic system absorbs and mixes with only as much water as is needed to complete the foaming and curing of components. A typical hydrophobic system expands ≤30-fold its initial volume. In the past, most hydrophobic foams were rigid and could end-up brittle if not formulated carefully. As a consequence for concrete-crack repairs and other types of sealing, contractors used hydrophilic systems. As technology improved, hydrophobic systems that were more flexible and stable were available to stop water and repair cracks. These systems bond well to concrete, do not shrink even in the absence of water, and can expand enough to fill voids economically.

Two-component hydrophobic systems can be formulated if one side contains catalysts and the other side contains reacting ingredients. These systems need very little water to initiate the foaming reaction. Typically, there is enough water present in a crack to initiate a reaction. Thus, unlike hydrophilic systems (in which water is mixed first with a hydrophilic system), hydrophobic systems do not react until already present in the crack. A hydrophobic system has time to permeate all parts of the crack before beginning to foam. A hydrophilic system usually foams before entering the crack, can clog-up the crack, push the water back into crevices, and cause leaks later.

A hydrophilic system acts like a sponge and absorbs as much water as is available in its surroundings during foaming. It reacts with the amount of water needed for the reaction and subsequently contains the excess water within its structure (as would a sponge). In dry conditions, some of the excess entrapped water can evaporate and result in foam shrinkage. A typical hydrophilic system is very flexible, has good adhesion to concrete, and bonds well to the sides of a concrete crack. This excellent adhesion and flexibility can minimise shrinkage in a crack. A typical hydrophilic system expands to 2–4-fold its starting volume. It is an excellent product if there is limited movement around a crack which is >3/8 inch and if its flexibility can be realised. Overall, hydrophobic systems that have good adhesion and flexibility are excellent for most repairs of basement concrete cracks, whereas hydrophilic systems are good if there is movement and water is always present.

5.3.1 Applications

The following are some of the basic effective and useful applications:
- Foundation cracks
- Floor and slab cracks
- Repair of crawl space
- As sealants
- Waterproofing of drainage systems

5.4 Polyisocyanurate foams

Polyisocyanurate foams can be made by a reaction of a polyisocyanate with a polyol composition mixture comprising a polyol, water, a trimerisation catalyst, a carbodiimide catalyst and an aromatic compound used as a chain extender. The amount of the aromatic compound used in the mixture is 0.8–10.0 wt% of the total mixture, whereas small amounts of heat stabilisers, cell stabilisers and pigments/dyes are added as needed.

5.4.1 Applications

Applications of polyisocyanurate foams are:
1. Heat-insulation panels
2. Heat-insulation materials
3. Carpet underlay

5.5 Soy–phosphate polyol rigid polyurethane foams

Refined soybean oils (SBO) contain ≈99% triglycerides with active sites amenable for chemical reactions. Therefore, SBO are promising bio-based resources as feedstock of polyols for manufacture of polyurethanes (PU). Soy–phosphate polyol (SPP) is made from SBO-derived epoxides with the presence of phosphate acid as a catalyst by acidolysis. Usually, acidolysis is carried out by mixing SBO-derived epoxides, phosphate acid, water and polar solvents. The reaction occurs instantly to produce clear, viscous, homogenous SPP with a low acid value and high functionality. In this reaction, SBO-derived epoxide can react with water directly to form diols because of their high reactivity through cleavage of the oxirane ring. Phosphoric acid catalyses ring-opening reactions and is chemically involved to become part of the polyol.

Distilled water, acting as a blowing agent, reacts with isocyanate to generate carbon dioxide (CO_2), which creates foaming of the reactant mixture to form a cellular structure. This is an important parameter that influences the properties and performance of rigid PUF. By varying the water content, increase or decrease in the density of the foam can be achieved. Strength increases with increases in the density of the foams. Increase in water content produces more CO_2, and results in foam with larger volume and lower density. Here, the foam morphology is influenced by the water content. Increased water content produces rigid PUF with thinner foam cell walls and larger foam cells, thereby changing the density and morphology of the foam. Another influential component in water-blown PU foaming using SBO polyols is isocyanate content. Tests have shown that the effect of the isocyanate index varies the compressive strength of SBO foams proportionately if the isocyanate index is >100, suggesting that the mechanical properties of SBO rigid foams can be modified by altering the amount of isocyanate in the foaming formulation.

5.5.1 Foaming method

I am providing guidelines for small-volume production, the principles of which can be adopted for large-scale production with slight modifications of formulation to suit sophisticated dispensing machines. A water-blown rigid PUF can be made on the basis of a two-component system. For example, 100 pbw of a standard polyol (Voranol 490) and 50 pbw of a soy–polyol are weighed and mixed together in a container. Other component B-side ingredients (water, catalysts and surfactants) are added into this mix. This combination is allowed to mix for ≈2 min at 3,450 rpm. This mixture is allowed to degas for ≈2 min. A pre-weighed component A (isocyanate) is added into this mix, and mixed for 10 s at the same speed. This reacting mixture is poured rapidly into a wooden mould of dimension 12 × 12 × 22 inches lined with release foil or waxed inner surfaces for easy release. The reacting foam rises slowly and sets at ambient temperature. Demoulding should occur when surface ceases to be tacky. By adjusting the water content in the formulation to 2–4%, different foam densities can be achieved.

5.5.2 Applications

One of the main uses of rigid PUF is insulation. Compared with other insulating material, PU rigid foam is highly competitive. There are five product-related advantages: low thermal conductivity; high mechanical and chemical properties at high and low temperatures; all major international fire-safety requirements can be met; the ability to form sandwich structures with various materials; and the new generation of PU is recyclable.

Rigid PUF perform well in most low-temperature insulations. Foam products with density 30–200 kg/m^3 can withstand temperatures down to 160 °C. Typical applications are: refrigerated vehicles; road and rail tankers; vessels for refrigerated cargo; pipelines; liquid gas tanks for liquefied petroleum gas; cryogenic wind tunnels.

5.6 Integral skin polyurethane foams

Integral skin PUF are specially produced moulded foams with aesthetic features for special applications. Basic integral skin foam (ISF) product consists of a soft cellular core with a thin or thick surface skin. The latter can be in almost any colour and depends on the end-application. The foam formulation can be a conventional one with polyether polyol, isocyanate and water as the blowing agent. However, producing a good skin layer is difficult because the CO_2 generated by the reaction between isocyanate and water at the mould surface is unlikely to condense satisfactorily. New technologies promote production of all water-blown ISF products with a new polypropylene glycol (PPG) which contains extremely small amounts of by-products and has a narrow molecular-weight distribution compared with conventional PPG. In this system, controlling the balance between the gelling reaction and blowing reaction of ISF and also obtaining very thin/light skins is easy.

PU ISF can be made in a range of hardness values starting from 30 up to 85 shore A, in densities of 350 up to 800 g/l. Colouring PUF is possible by addition of pigments with good dispersion properties. In general, it is best to add pigments into the polyol base and to mix thoroughly before the other ingredients are brought on-stream. ISF products can be made using an 'open-mould' or 'closed-mould' concept. Inserts of any type can be incorporated into the foam during processing. PU ISF can be used in the production of three-dimensional (3D) products irrespective of whether they are fitted with a metal and/or wooden inlay. These inlays: i) are integrated during moulding to provide a final product with better dimensional stability; and ii) provide possibilities to connect/assemble to another end-product. By using special additives different properties can be achieved in the foam according to end-applications.

ISF or 'self-skinned' foam is a natural occurrence during injection moulding, and occurs when the resin interfaces with the high temperature of the aluminium mould. The high temperature 'cooks' the resin so completely that the size of foam cells is reduced to microns, creating an extremely dense version of the foam on the outside (skin thickness) with seemingly no cell structure. The mould temperature plays an important part in formation of the thickness of the skin. The foam cells that reside below the skin are well protected by the skin, thus creating a durable protective layer. The strong surface skin provides protection against scratching, wear, tear and abrasion, which improves durability in a rough physical environment. Usually, it takes many hundreds of hours of constant abrasion to remove the skin and expose the cell structure within. Products that require long-lasting usability can benefit from these tough ISF. Whether it is an outdoor, indoor, commercial or industrial application, the unique functions of ISF play a major part.

5.6.1 Applications

Some of popular end-uses are automobile fixtures, armrests, headrests, backrests, seats, safety bars (playgrounds), toilet seats, backpacks, kayak rack, and medical devices.

5.7 Convoluted and contoured cellular foams

'Convolution' is a maximum yield process which alters the surface of a foam on a customisable basis to provide enhanced benefits based mainly around comfort and pressure distribution. The foam is made in a conventional way and 'foam buns' are cut into desired thicknesses. The density of the foam determines the soft/hard quality of the final convoluted sheets from which different sizes and shapes as required by customers are cut to suit end-applications. Convolution is a compression process, so loss of height of ≈10% should be factored in when thicknesses of the final product are determined.

Convolution alters the foam surface by pattern, size of pattern, depth of cut and spacing between patterns.

Contoured patterns are cut from solid foam buns or blocks. A producer can use a single or multiple hotwire system to cut foams (discouraged due to fire hazard), but the conventional way is by using band knives. There are many types of band knives, and semi-automatic and fully automatic cutting machines are available.

Some of the specialised foam-cutting machines are:
- Bevel cutting machines
- Peeling machines
- Profile-cutting machines
- Edge-profiling machines
- Contour-cutting machines
- Convoluting machines
- Sheet-cutting machines

5.7.1 Applications

Foams can be used for many applications:
- Bedding
- Sound insulation
- Packaging
- Enhanced air circulation
- Healthcare applications
- Pressure distribution for comfort
- Overlays
- Support bases for mattresses
- Mattresses
- Luxury furniture

5.8 Reticulated foams

Reticulated foams are produced from sheets from a foam block by an explosion process. That is, a foam block of controlled pore size is placed in a pressure vessel with reactive gases which, when ignited, blow out ≤98% of the cell walls of the foam. The structure thus formed is a dodecahedron. This is why the cell size before 'blowing' is the single most important characteristic of reticulate foams (closely followed by the consistency of cell dimensions). These characteristics determine the nature and working parameters of the filter itself.

Reticulated PUF is a versatile open-cell material that is lightweight, with low odour and high resistance to mildew. The porosity of reticulated foams is vital when designing a custom component. A wide range of these foams with different porosities are available. Foam technology for the manufacture of these special foams involves manipulation of thousands of polymer bubbles (cells) of precisely controlled sizes. Reticulation is a post-operation of conventional water-blown PUF that removes the windows (walls) of the cells. The cells that make up a foam can have several variations that can also be controlled precisely. Different foams have different cell structures and characteristics, but foams from the same material family can be made with vastly different densities and firmness specifications that can affect their performance greatly.

There are two basic methods of reticulation: thermal (also called 'zapping') and chemical (also called 'quenching'). Reticulated foams can act as filters. Filter foam is a reticulated PUF adapted specially to filtration of air and liquid in a range of controlled cells of 10–100 pores/inch. Filter foam with a medium porosity 45 pores/inch works as a depth-loading filter as opposed to a surface-loading filter, thereby trapping dust particles within the cell structure. Reticulation leaves behind a skeletal structure of the foam (97% void volume), thereby giving it a high surface area for collection of dust particles. With their homogenous and uniform cell structure, reticulated foams can be engineered for pressure drop and filtering efficiency by changing their pore size. For example, a coarse porosity of 10 pores/inch is effective as a sound attenuator, scrubber pad, and washable filtration media for air conditioners, furnaces, small engines and automobile air cleaners.

If reticulated foams are compressed, the material takes on a new set of properties ideal for other applications requiring high void volumes, uniform porosity, non-directional characteristics, exceptional breathability and uniform texture.

Reticulated foams made from PUF were developed for increased hydrostatic stability. Polyether foam is a flexible compressed open-cell type of PUF that is smooth in texture. Polyether foam is manufactured by mixing polyether polyols, a catalyst, a surfactant and an isocyanate with water as the blowing agent to produce a free-rising foamy bun that solidifies within minutes.

Reticulated foams made from polyester urethane foams have a 3D skeletal strand structure that minimises the possibility of open channels and provides excellent filtration properties. Polyester foam is flexible, open-cell, porous and has a uniform cell structure. It has an evenly spaced cell structure with a high proportion of closed-cells or 'windows'. The large amount of windows makes polyester foam ideal for sealing applications because it prohibits the flow of air naturally. The uniform cell structure inherent in polyester foam also makes it ideal for the reticulation process. Polyester foam can be compressed easily to form sheets with a fixed desirable thickness. Additives used during the manufacturing process can transform the foam properties to make it flame-retardant, antimicrobial, antistatic and conductive. These filter foams are used widely for filtration of air and liquids.

5.8.1 Applications

These foams have many end-uses in domestic, industrial and engineering applications:
- Humidifier pads
- Filtration of air, water or dust
- Scrubbers
- Ceramic filters
- Military and medical products
- Costumes
- Automobile filters
- Bacteria filters
- Speaker grills
- Odourising wicks
- Facemask pads

5.9 Sulfone cellular foams

Sulfone cellular foams are made from sulfone polymers and mixtures of non-sulfone polymers. They are produced using aqueous blowing agents comprising water as the only blowing agent or mixtures of water with another blowing agent. I will deal with cellular foams produced with water as a blowing agent for the production of low-density polysulfone foams. Sulfone polymeric materials are well-known for their useful characteristics as engineering thermoplastics for high-temperature applications. These foams are difficult to make because high temperatures are required to produce a flowable gel from sulfone polymeric materials with the usual blowing agents.

The general principle in producing sulfone polymer foams is use of water as the blowing agent. First, a mixture comprising one or more sulfone polymers and water as a blowing agent under conditions of sufficient heat and pressure is converted into a flowable gel. Then, the pressure is released to make the gel into a cellular mass.

5.9.1 Applications

In addition to the conventional uses of cellular foams, sulfone cellular foams are used for special engineering applications where high temperatures are involved.

5.10 Additives for cellular foams

The polymer industry has experienced several changes throughout the years, starting with the discovery of natural polymers and evolving into more advanced polymers and even more specialised polymers. The plastics 'revolution' has virtually replaced the used of traditional conventional materials, and there is no limit to the use of plastics for any end-application. Polymeric composites with biomasses can be used to produce 'polymer lumber', which is considered to be an ideal substitute for natural wood and even superior in properties. Due to their natural high densities, additives play a key part in enhancing its properties and density reduction through cellular structures.

Polymer additives are an important area of innovation for the plastics industry. An additive is a material that is added to a polymer melt to enhance processability, performance and aesthetic appearance. Additives such as blowing agents play a key part in production of cellular foams. Most additives used to be based on petroleum but, due to environmental concerns, industrial and commercial worlds are looking for polymers from natural sources, and suitable additives are also needed.

Natural and synthetic polymers rely on additives and special additives for processing. Manufacturers of resins are looking for newer and cost-effective additives to produce polymers and blends with improved characteristics, with cost-effectiveness also being a key factor.

5.10.1 Types of additives

Additives can be categorised into: stabilisers, processing aids, plasticisers, anti-static agents, blowing agents, fillers, coupling agents, antibacterial agents, catalysts, surfactants, cell openers and colourants. These additives, alone or in combination, are vital for the polymer industry.

5.10.2 Stabilisers

Stabilisers can be categorised broadly into four types. The first type is antioxidants, which can protect a material during processing and extend its longevity. They are used to prevent polymer degradation that can result in loss of strength, flexibility, thermal stability and colour. Antioxidants eliminate oxidation during and after processing if materials are exposed to an energy source.

A second category of stabilisers are those that help materials to withstand UV light. UV radiation damages the chemical bonds of polymeric materials. Therefore, it is essential to add stabilisers to materials exposed to extended periods of UV action (especially for products used outdoors). These stabilisers absorb high-energy UV radiation and then release it at lower energy level that is less harmful for the polymer. For example, titanium dioxide has a high refractive index that enhances long-term stability and protects against material discolouration.

A third category of stabilisers are heat stabilisers, which prevent thermal breakdown of materials and preserve aesthetic properties. They eliminate chemical decomposition during processing.

The final category of stabilisers is flame retardants. Many polymers are flammable in their pure form and this becomes a liability (especially for the building industry) unless the materials can be stabilised. Flame retardants slow down combustion or create a new reaction that produces less heat and dies out. These additives can be processed easily and have no impact on the other physical properties of materials.

5.10.3 Lubricants

Lubricants are used to improve flow and processing properties. There are various lubricants with varying properties that can greatly improve mixing and provide smooth flow during processing. An 'internal lubricant' is a type of additive that modifies the viscosity of a material. Choosing the correct grade to suit a particular material and process is important, and lubricants have value-added performance because they can also modify other properties. They can be cost-effective by achieving multiple goals with the same lubricant.

'Slip additives' are a type of internal lubricant that create better processability by reducing the internal friction and tackiness of polymers or a mass. They are used commonly for film manufacture because they help film layers to slide over each other, which can be very useful in high-speed operations.

Anti-blocking additives are another type of internal lubricant that can improve processability during production. Many types of anti-blocking additives can provide diverse offerings for formulations, and choosing the correct ones for cellular foams (where necessary) is important. For example, expensive synthetic silica additives give high clarity. Alternatively, calcium carbonate is a simple effective additive and, because most cellular foams are likely to have this additive already in a formulation, additional ones may not be required. If these two internal lubricants are used in combination, anti-blocking reduces the effectiveness of a slip agent.

5.10.4 Plasticisers

Plasticisers are additives that increase the plasticity or flexibility of a plastic material. A plasticiser softens the final product, thus increasing movement and durability. Plasticisers embed themselves between polymer chains and push chains further apart, resulting in a more flexible plastic but causing a loss of strength and hardness. Plasticisers are used mostly with polyvinyl chloride (PVC) because without this additive they are too rigid to be processed. For example, raw unplasticised PVC is brittle and rigid and requires additives before processing. Plasticisers are prone to UV action and, if PVC pipes are exposed to the atmosphere for long periods, they become brittle due to evaporation of the plasticiser in the material.

5.10.5 Blowing agents

Blowing agents are additives that decompose to form a gas that produces cellular structures in polymers. Blowing agents are used to create foams and expandable materials that are lightweight and provide comfort as well as protection against heat or shock. Most blowing agents are based on petroleum, expensive, and result in harmful emissions; some of the oldest blowing agents are banned. New water-blowing methods are being developed.

5.10.6 Fillers

Fillers are added to polymers and formulations primarily to reduce costs, improve properties, increase density and increase compression strength. Polymers use less expensive fillers to replace some of the volume of more expensive materials. Fillers can improve processing, abrasion resistance, density control, dimensional/thermal stability and optical effects. Common fillers can be wood flour, silica, glass, or clay talc, but the most common filler used for cellular foam is calcium carbonate (as a fine powder).

5.10.7 Coupling agents

The primary purpose of using coupling agents is to increase interaction between polymers and fillers or other material. They create chemical links between molecules to improve bonding. When the coupling agent or a system bonds to a polymer, they can enhance the adhesion between two or more materials. Coupling agents promote bonding, so they can be used to encourage materials that are normally incompatible to bond together. Compatibiliser agents can also be used for this purpose. This function can be useful when trying to create new polymer blends and for using recycled material.

5.10.8 Antibacterial additives

Antibacterial additives are used to create resistance to, and counter the actions of, microorganisms in foams so that polymeric foams are protected from bacterial growth. Antibacterial additives interfere with the metabolism of microorganisms and block enzyme systems. To be effective the additive must migrate to the surface of the material so that it can interact with microorganisms.

5.10.9 Catalysts

A range of catalysts is available. Some catalysts promote and accelerate chemical reactions whereas others slow-down reactions. Tertiary amines energise, accelerate and control the rate of the reaction between water and isocyanate, whereas tin salts are specific for the reaction between polyol and isocyanates in polymer foaming. The tin catalyst used almost universally is stannous octoate (and is often referred to as the 'tin catalyst'). Accurate weighing of catalysts is important because only very small quantities are required.

5.10.10 Surfactants

A surfactant is essential for control of the foaming process. In PU manufacturing processes, the most commonly used surfactant is a silicone surfactant. Its two main functions are to: i) assist mixing of the components to form a homogenous liquid; and ii) control and stabilise the bubbles/cell structure in the foam during expansion and to prevent collapse before the liquid phase has polymerised.

5.10.11 Colouring additives

Most cellular foams are colourless or slightly yellow. Colouring is done by dyes, pigments or masterbatches to obtain the desired colour of the end product. Colour changing is essential in polymer foams to counter environmental issues such as temperature, weather, UV action, or to enhance aesthetic values, which are an essential part of plastics. Some large-volume manufacturers of foams enable many types of foam blocks with different properties to be made, 'density colour' their blocks, and use these codes to identify them before fabrication. Most producers of colouring products have codes that adhere to acceptable international standards to ensure safety (especially in food applications). Sometimes, a foam manufacturer may need a colouring medium with more than one function and it is best to be guided by them. As a general rule, colours coded from 0 to 8 denote a gradual increase in quality, with 8 being the best for colour fastness. A basic yellow additive colour will counter UV action which discolours and degrade foams.

5.10.12 Cell openers

In some PUF, it is necessary to add a cell opener to prevent cells from shrinking upon cooling. This is more prevalent in polymer cellular foams. Additives used for inducing cell opening and stabilising them are silicone-based waxes, finely divided solids, paraffin oil, long-chain fatty acids, and certain polyether polyols made using a high concentration of ethylene oxide.

5.10.13 Dispensing equipment for additives

Additives come in different forms: powders, liquids, solids or masterbatches. Some colouring masterbatches may contain more than one additive. There are a few vital pieces of equipment and machinery that enable accurate dispensing (which is needed because the amounts of additives are usually small). Volumetric/gravimetric feeders and blenders are the most widely used equipment that measures/weighs and dispenses additives into polymer blends.

Volumetric devices measure volume that passes through a metering disc, whereas gravimetric devices measure and control the weight of additives dispersed over a given time. Gravimetric devices are recommended for operations if two or more additives are being dispensed at a machine throat or in a blending system. The blending system can be a dosing or mixing unit, which divides the main component into different streams so these streams can combine with additives at the feed inlet to produce a homogenous mixture. Another important piece of equipment is level sensors because they can measure insufficient additive material to prevent unnecessary downtime.

A small producer of foam may opt for a simple manual operation of weighing using electronic devises for each additive but must be very accurate because small quantities are involved.

Bibliography

1. *Reticulated Polyurethane Foam*, UFP Technologies, Newburyport, MA, USA. http://www.ufpt.com/materials/foam/reticulated-polyurethane-foam.html.
2. J. Fox in *Analysis of Polymer Additives*, University of Florida, Gainesville, FL, USA, 2008.
3. *Integral Skin Foam or Self-skinned Foam*, Peritek B.V., The Netherlands.
4. H. Wada and H. Fukuda, *Journal of Cellular Plastics*, 2009, **45**, 4, 293.
5. H. Fan, A. Tekeei, G.J. Suppes and F-H. Hsieh, *International Journal of Polymer Science*, 2012, Article ID: 907408. http://dx.doi.org/10.1155/2012/907049.
6. *Use of Hydrophobic and Hydrophilic Polyurethane Foams for Crack Injection*, Emecole Inc., Romeoville, IL, USA.
7. *Spray Foam Urethane Technology*, Urethane Technology Company Inc., New York, NY, USA.
8. *Reticulated Foam*, Vita, Toronto, Canada. http://www.engineeredfoam.com/reticulatedfoams.php.

6 Raw materials, storage of materials and basic safety factors

6.1 Raw materials

The subject of cellular polymers covers a wide range of foams. Hence, this chapter will deal only with raw materials relevant to water-blown polyurethanes (PU) and polystyrene, which form ≈75% of the cellular market. For water-blown expandable polystyrene, the extra water is incorporated into the styrene monomer during polymerisation. However, urethane production and their foaming is a more complex matter involving several raw-material components and the requirement of appropriate formulation.

Raw materials for making cellular polyurethane foam (PUF) include a range of polyols, graft polyols, natural oil polyols (NOP), isocyanates, diphenyl methane diisocyanate (MDI), water, surfactants, catalysts, cell openers, pigments, dyes and other additives. These must be formulated in different combinations to achieve foams with desired properties to suit end-applications. The two basic foam types are open cell (flexible) and closed cell (semi-rigid or rigid).

These raw materials are combined in 'systems' rather than a common standard formula. They are formulated to achieve desired properties to suit end-applications and to suit a particular customer or for a common marketplace. The two most important basic final properties to be achieved are the density of the foam and, in the case of comfort, the support factor. All raw-material systems or formulae are designed around these vital aspects. These raw materials or components can be purchased separately as systems or two components: component A (isocyanate) and component B (polyol + additives). Purchasing separately yields cost savings and the ability to formulate different densities and properties, but the foam producer must have sufficient skills to formulate correctly. If such skills are absent, then wastage, hazards and poor-quality blends due to mistakes in weighing different components (especially additives, which are in very small quantities) can result.

Alternatively, a foam producer could request professional blending companies to supply blended systems in the form of two- or three-components, which are easy to store, handle and process. However, this results in extra costs. The market prices for viscoelastic or memory foams are high, and one could expect larger profit margins, but this may be a better and safer method for securing raw-material supplies. Density ranges are limited, so a foam producer could opt for purchases on a 'just-in-time' basis. Individual raw materials are available in steel drums and smaller containers, but blended systems may come in drums, totes or tankers (bulk). The last option is for large-volume foam manufacturers and has the advantages of cheaper costs and ensuring availability of raw materials for production at all times, thereby avoiding delays in obtaining raw materials due to various reasons. However, the layout for

https://doi.org/10.1515/9783110643121-006

such a system, in which the bulk materials are pumped directly from tankers into large receiving tanks, incurs heavy investment. Also, safety factors must be taken into account.

6.1.1 Polyols

A polyol is an alcohol containing multiple –OH functional groups available for organic reactions. They are high-molecular-weight (MW) materials manufactured from an initiator and monomeric building blocks. They are classified into 'polyether polyols' and 'polyester polyols'. The former are made by the reaction of epoxides (oxiranes) with active hydrogen containing a starter compound. An epoxide is an ether with a cyclic structure with 3-ring atoms. Polymerisation of an epoxide gives a polyether, whereas polyester polyols are made by polycondensation of multifunctional carboxylic acids and OH compounds. Polyether polyols offer better technical and commercial advantages (e.g., low cost, ease of handling, flexibility and hydraulic stability) over polyester polyols.

Polyols can be classified further according to their texture (flexible or rigid) depending on their MW and functionality. Flexible polyols have MW from 2,000 to 10,000 with the hydroxyl number (OH#) ranging from 18 to 56. Rigid polyols have MW from 250 to 700. Polyols with MW from 700 to 2,000 and OH# ranging from 60 to 280 are used to add stiffness as well as solubility in both low-MW glycols and high-MW polyols.

Graft polyether polyols contain copolymerised styrene and acrylonitriles. They are designed to maximise the load-bearing properties of PUF. Some of the new grades of graft polyols have solids in the range 10–45% in the production of slab-stock (large foam blocks made by a continuous process).

NOP (also termed 'bio-polyols') are polyols derived from vegetable oils. NOP have similar sources and applications but the materials themselves can be quite different. All are clear liquids, ranging from colourless to medium yellow. Their viscosity varies and is usually a function of the MW and average number of OH groups per molecule. Higher viscosities have higher MW and higher OH contents. Odour is a significant property, which differs from NOP to NOP and quite similar chemically to their parent vegetable oils and, as such, they are prone to becoming rancid. This is one of the disadvantages of bio-polyols and remedial action must be taken. Odour is undesirable in the NOP themselves but, more importantly, in the final materials made from them.

There are a limited number of naturally occurring vegetable oils (triglycerides) that contain the unreacted OH groups that account for the name and reactivity of these polyols. Castor oil is the only NOP available commercially that is produced directly from a plant source. All other NOP require chemical modification of the oils available directly from plants.

Due to years of experience and research, developers of chemical products can make oils and fats into products that have better value in the chemical industry (rather than just as simple lubricants or additives). These activities have resulted in the availability of a brand called ProloOil polyols. They are categorised into two general groups (A and B polyols) each with varying hydroxyl values (OHV). These OHV represent the reactivity of the polyols as functional groups. It is these OH groups that react with isocyanates and MDI to form urethane monomers and, hence, PU. 'A' polyols are slightly more reactive than 'B' polyols. 'B' polyols are more branched than 'A' polyols. 'B' polyols are used for flexible foams and 'A' polyols used for rigid foams. Thus, a 15A polyol is less reactive than a 20A polyol because it has fewer OH groups.

'Green' polyols are of two types: polyether polyols and polyester polyols. Bio-polyols are a category of polyols synthesised from natural oils (e.g., soya, castor, rapeseed, corn, canola, and palm). Green polyols are synthesised by recycling polyethylene terephthalate and PU wastes. Both categories are used in applications such as flexible and rigid foams, coatings, adhesives, sealants, as well as other applications in construction, transport, carpet underlays, packaging, furniture, and bedding. Bio-polyols and green polyols are being used to substitute petroleum-based conventional polyols owing to advantages such as:

- Availability of raw materials
- Increasing prices of crude oil
- Lower carbon footprint

6.1.2 Isocyanates

Organic compounds that contain an isocyanate group are called 'isocyanates'. An isocyanate that has two isocyanate groups is known as a 'diisocyanate'. Isocyanates with two or more functional groups are required for the formation of PU polymers. Diisocyantes are classified into 'aromatic', 'aliphatic' and 'cycloaliphatic'. Aromatic isocyanates are the most widely produced isocyanates for two main reasons: i) aromatically linked isocyanate groups are much more reactive than aliphatic ones; and ii) aromatic isocyanates are economical to use. Aliphatics are used only if special properties are required for the final product. The two most important commercial aromatic isocyanates are toluene diisocyanate (TDI) and MDI.

The most common isocyanate used in the manufacture of cellular PUF is TDI. The latter is a colourless to pale-yellow liquid, toxic, with a density of 1.21 g/cm^3 and a melting point of 21.80 °C. MDI is also used for cellular foams but more for integral skin and moulded products. Mixtures of 2,4-TDI and 2,6-TDI are the isocyanates of choice for the production of flexible cellular PUF. TDI are low-cost, high-quality products that allow foam manufacturers to produce many types of flexible foams with a wide range of physical properties. TDI is available in 2,4-TDI:2,6-TDI isomer mixtures of 80:20 or 65:35, as well as a pure 2,4-TDI isomer. However, the reactivity of the system and resulting foam properties can be modified using blends of the various isomer ratio mixtures.

6.1.3 Catalysts

Virtually all commercial polymer foams are made with the aid of at least one catalyst. A catalyst is a substance that alters the rate of a chemical reaction but remains unchanged. A 'positive catalyst' accelerates a reaction. A 'negative' catalyst slows down a reaction. All catalysts in polymer foaming are positive catalysts. Most foaming systems contain a catalyst system to achieve optimum balance between the chain propagation and blowing reaction. Rates of formation of polymers and gas are controlled so that the gas is entrapped efficiently in the gelling polymer and cell walls become strong without collapsing or shrinking. Another important part played by catalysts is that they ensure full 'cure' in finished foams.

The type and concentration of catalysts can be selected to satisfy process requirements such as cream time, rise time, profile, gel time and final cure of the outer skin. Two broad classes of catalysts are used in polymer foaming: amine and organometallic (commonly called 'tin catalysts').

Amine catalysts such as triethylenediamine (TEDA) or, for example, Dabco and dimethylethanolamine, are used most commonly for cellular foams. Tertiary amines are selected on the basis of whether they drive the gelling reaction or blowing reaction. Most tertiary amines drive both reactions to some extent, so they are selected further based on how much they favour one reaction over the other. For example, tetramethyl-butanediamine drives the gelling reaction better than the blowing reaction. TEDA drives the blowing reaction better than the gelling reaction.

An organometallic compound is a compound containing at least one metal-to-carbon bond in which the carbon is part of an organic group. Organometallic compounds based on mercury, lead, tin, bismuth and zinc can be used for cellular foams. In PUF, the polymer foam making reaction (gelling) between the isocyanate and polyol is promoted best by tin catalysts, which is generally the preferred one.

Water, surfactants and catalysts have been discussed in earlier chapters also, but these being of utmost importance, I present below some of their basic functions for production of water-blown cellular foams.

Polycat SA-102 is a catalyst in a delayed-action tertiary amine that is activated by heat and strongly promotes the urethane (polyol isocyanate) reaction in microcellular, rigid, integral skin and sealant foams.

Polycat 140 is a low-emission amine catalyst for low-density, water blown, open-cell spray PUF. This is a reactive catalyst and better for the water–isocyanate blowing reaction. When used with Polycat 141 or Polycat 142, it can replace standard blowing catalysts such as Dabco BL-11 and Dabco BL-19.

Polycat 141 is a low-emission amine co-catalyst for low-density, water-blown, open-cell spray PUF. If used in combination with a base catalyst such as Polycat 140, it improves the water–isocyanate blowing reaction.

Polycat 15 is a non-emissive balanced amine that has slight selectivity towards the urea (water–isocyanate) reaction. Due to its reactive hydrogen, it reacts readily into the polymer matrix. It also improves the surface cure in flexible moulded systems and can be used in rigid PU systems (in which a smooth reaction profile is desired).

Polycat 17 is a non-emissive balanced amine catalyst that has slight selectivity towards the urethane (polyol isocyanate) reaction. Due to its reactive –OH group, it reacts readily with the polymer matrix. It can also be used for low-density, semi-flexible PUF and low-density rigid PUF for packaging.

Polycat 201 is a new, novel non-emissive, and low-odour balanced amine catalyst providing excellent shelf-life, stability and reactivity performance for low-water-blown systems.

6.1.4 Water

The polymerisation reaction between the polyol and isocyanates produces a solid, extremely high-density and rigid mass of PU but certainly no foam. What gives the PUF its low mass-to-volume ratio is expansion of gases such as carbon dioxide (CO_2), which can also be introduced into a reaction mix but CO_2 generated endogenously is not only more economical but also the blowing phase is easier to control. The function of the water in the formulation is to react with the isocyanate to produce CO_2, which is exothermic (heat giving) and also produces urea. A combination of the heat generated from the polymerisation process and that from the water–isocyanate reaction makes CO_2 expand within and gel the PU. In most flexible slabstock foam, water is the only or primary blowing agent.

In general, ordinary water sources (e.g., from the tap) are good enough. However, water with high concentrations of dissolved or suspended metals are not suitable unless they are treated because unwanted metals may interfere with catalytic reactions.

6.1.5 Surfactants

Silicones make an ideal product for creating a stable homogenous blend of 'un-blendable' components such as oil and water. The most used silicone surfactants are organo-modified (non-ionic) branched silicon polymers with many 'heads' and 'tails' (hydrophilic and hydrophobic) which makes them more efficient than most other surfactants and most cost-effective to use. In foams, silicone surfactants not only emulsify, disperse and help high-speed blending of all raw-material components but have an even more important role of stabilising foam cells during and after blowing phase. Blowing and expansion of gases takes place almost at the gelling stage of the PUF, so the polymer matrix is not yet strong enough to support cell walls against the motive force of expanding gases.

If not for the presence of the surfactant, gases rupture cell walls and escape without expanding cells, and the blowing phase is not successful. Surfactant molecules engulf foam cells, aligning themselves at the interface between PU cells and the air space within cells. As a result, the cell walls are given enough strength to contain the pressure of the expanding gases without rupturing. This process is called the 'stabilisation phase' and silicone surfactants in foaming are known as 'foam stabilisers'. There are different grades of silicone surfactants, and a foam producer selects the most suitable ones in keeping with the types of cellular foams being produced.

6.1.6 Additives

Apart from producing standard foams, foams with special and different qualities may be required in cellular foams. Because of the end-uses of foams (e.g., comfort, industrial, automotive, and domestic), foams may need to have special properties: load bearing, fire retardancy, resilience, fluid management (filtration), absorption. Additives are needed to achieve these special qualities. The three broad areas for specialty properties are: chemical, physical, mechanical, and space travel. Usually, a foam producer may opt to use more than one additive in the form of a system to achieve the end objective. These additives are discussed in more detail in Chapter 5.

6.1.7 Fillers

Usually, fillers come as finely divided inert inorganic material added to foam formulations to increase density and load bearing, and to reduce cost. A wide range of fillers are available but only inorganic calcium carbonates are used commonly in production of cellular foam. Knowing the water content of the filler used is important to avoid going beyond the water threshold for a particular formula.

6.1.8 Colourants

Most foam producers use colour codes to identify different qualities of foams. Yellow can counter the action of ultraviolet (UV) light, which discolours and degrades a foam if exposed for long periods. Coloured foams also are pleasing to the eye and have good aesthetic values. Colouring can be done by blending very small quantities of pigments or dyes in the polyol content. Typical colouring inorganic agents include titanium dioxide, iron oxides and chromium oxide.

6.1.9 Need for colouring

In general, most plastics resins are colourless, opaque or translucent irrespective of whether they come in the form of pellets, beads, powders, liquids or any other form. Manufacturers of plastics resins and polymers can colour them during polymerisation (joining of single mers) or offering suitable dyes, pigments or masterbatches to colour them as needed by the producers of plastics products. In large-volume production of foam, in which many types of foams are made under one roof (e.g., different densities, physical properties and specialty end-applications), foam producers may opt for colouring the foams for easy identification, aesthetic values or according to customer requests. UV rays (sunlight) tend to discolour and degrade foams, so producers colour their foams yellow.

6.1.10 Colour wheel

Most foam manufacturers have small in-house laboratories in which they can carry out experiments and pre-production foam trials ('box-tests'). These facilities are to make the desired colour blends provided the basic knowledge is known. For colouring foams, the best method is to mix it into the polyol before the other components are brought on-stream.

The 'colour wheel' (or 'colour circle') is a basic tool for combining colours. The colour wheel has been designed so that virtually any colour you choose from it looks good together with another colour. Over the years, many variations of this design have been made. However, the most common version is the original wheel of 12 colours consisting of primary, secondary and tertiary colours. In practice, it is possible to produce pleasing colours ('colour harmonics') and they consist of more than two colours. They can be classified as 'warm' and 'cool' colours, tints, shades and tones. However, basic colours should be more than sufficient for the colouring of general-purpose cellular foams.

6.1.11 Primary colours

According to the red, yellow and blue (RYB) colour model system, primary colours are red, yellow and blue. I suggest that two other colours, black and white, should be added and this forms a more realistic scenario with five colours instead of three. Primary colours mean that they cannot be achieved by mixing of any other colours. These basic colours can be combined to form any colour desired (e.g., yellow + blue = green; red + white = pink; and red + black = brown).

These additives are mentioned specifically because of their importance in application areas such as the building industry, automotive and industrial. Low-density foams with open-cell structures have a large surface area and high permeability for ignition sources and oxygen. Flame retardants are added to reduce the possibility of flammability and for non-promotion of fire. Selection of suitable additive/additives depends on the degree of flammability, initial ignitability, burning rate and smoke evolution. The most widely used flame retardants are chlorinated phosphate esters, chlorinated paraffin and melamine powders.

6.1.12 Dyes, pigments and masterbatches

For colouring foams, a foam producer can use dyes, pigments or masterbatches. In general, manufacturers of colouring products make dyes as powders, pigments as pastes, and masterbatches as solids in polymers and also as liquids. Plastics injection moulders, blow moulders, extrusion processers will prefer dyes, pigments and masterbatches in solid forms, but a foam manufacturer will opt for these in liquid form because all chemicals used in foams are liquids. The best option would be colourants in polymer carriers that are compatible with the polyols being used. In foam productions, colourants are first mixed into the polyol content before the other components are brought on-stream.

6.2 Material storage

Different suppliers of raw materials may have different types of containers. However, the most popular and standard shipping containers are:
- *Small plastic or metal containers*: most additives (except fillers) are liquids and required in very small quantities, so they are supplied in plastic containers (e.g., in 1–5 kg) whereas fillers come in larger (25 kg) packs in paper bags.
- *Steel drums*: each drum has two bungs (lids), one small and one large, with drums painted in colour for easy identification. General codes used are: red (isocyanates), blue (polyols) and green (others). Before removing the contents through the larger opening, it is advisable to open the smaller bung/lid slowly and let-off the built-up pressure inside. Unless the contents are transferred into larger holding tanks, the contents inside must be stirred for ≥10 min using a long-handled stirrer with a small 'disc' at one end to obtain a homogenous liquid.
- *Totes* are bulk packs available in plastic, steel, or plastic containers in a cardboard box. The strength of the container is important due to the immense weight of the contents. It is advisable to stir the contents for a short time before connecting the container to the foam-dispensing machine directly.
- *Tankers*: some large-volume foam manufacturers opt for supplies in bulk, which are delivered in tankers that can be pumped directly into very large

holding tanks (5,000–20,000 l). These systems have automatic temperature controllers and moisture-seeping barriers to maintain the contents in prime condition. These systems are connected directly to continuous foaming lines, and measured quantities can be directly pumped to the mixing head.

It is advisable to check the delivery papers carefully, especially the certificate of analysis, which gives information about the chemical make-up of each material (e.g., NCO content for isocyanates, and OH# for polyols). This information is very valuable later when calculating mixing ratios. If in-house laboratory services are available, a few random samples of the material can be checked for water contamination and other possible negatives. Each chemical must be pumped into the correct holding tank and, if mistakes are made, the results (and financial losses) could be dreadful. Recommended storage temperatures are 23.9–32.2 °C.

Materials in bulk storage must be treated with extreme care because they hold large quantities of chemicals. Improper storage conditions can result in fire hazards and financial losses (or both). Some basic requirements for good storage are:

- *Gas blanket*: the tanks holding the isocyanates must have an empty space between the top of the liquid level and inside top of the tank. This space should be filled with dry air or an inert gas (e.g., nitrogen) to prevent atmospheric water vapour reacting with the chemical inside.
- *Temperature control*: usually, bulk storage tanks are connected directly to foam-dispensing machines. Thus, the temperature of the chemicals are important as well as the pumps must be in good order to ensure the correct amount of chemical (as programmed) is metered to the mixing head.
- *Recirculation*: there should be a recirculation system inside a tank to ensure that the chemical remains in the fluid state as well as to maintain the temperature desired. If the material is allowed to stand, there is a possibility of minute 'freezing' and formation of tiny particles which clog the lines and pumping system.
- *Pre-mixing*: polyols may be blends or contain additives (especially in a two-component system) and, regardless of their containers, contents should be mixed thoroughly for ≥10 min before use to prevent 'settling' of additives at the bottom.

Most raw materials used for production of cellular foams are hazardous chemicals. Hence, good and safe storage of material starts with planning of a suitable building. The materials are flammable, corrosive and harmful, so appropriate storage and handling is crucial. This is stated not to create alarm but to stress the importance primarily to the personnel and overall operation. By adhering to well-established safety procedures, a company (or even a small foam manufacturer) can avoid unwanted accidents or incidents and ensure a smooth and profitable operation.

Large-volume producers may opt to house the raw materials, production areas, fabrication area, and finished products in different buildings to maximise safety. If so, each building should be separated by a few feet. An entrepreneur or a small-volume producer must house a complete foam operation in one building. If so, an appropriate layout could be: delivery of raw materials by a rear entrance and storage at the back of the factory building; machinery and production area; post-production area for post-curing; easy access from the post-production area on a 'first-in, first-out' system; fabrication and quality-control area; storage for finished goods; administration offices at the front of the building. A foam producer must consider having an in-house laboratory (even on a small scale), tool room, rest rooms, and workers' room.

At all times, a smooth-flow layout with easy access to each section must be ensured. Good ventilation, effective exhaust systems, safety wear, periodic information/training sessions for workers, and safe operating systems ensure safe, trouble-free and productive foam production. An in-house recycling operation greatly enhances the chance to keep the operation under control and provide a 'kickback' for profits through sale or in-house use of compacted foam wastes.

6.3 Safety factors

Safety factors are a must in any industrial operation and particularly in a foam-manufacturing operation. Extra effort is needed because the raw materials being used are toxic and hazardous.

6.3.1 Air

Fumes are generated from foaming operations and post-cure of large foam blocks. Exhaust systems are supposed to negate this effect, but there will be some pollution of air within the building. General recommendations are to get assistance from professional companies to carry out periodic tests to ensure air pollution is within an acceptable limit.

6.3.2 Production operation

All personnel must be protected from air pollution and from coming into direct contact with chemicals. The two most hazardous chemicals are the polyols and isocyanates. For large-scale foam producers, the hazards are less because the chemicals are contained in large vessels and pumped directly to machines. The smaller moulder must handle the chemicals manually or on a semi-automatic system and could come into direct contact with these two chemicals. Solvents are not envisaged in water-blown formulations, so a highly flammable hazard is eliminated. If direct contact is made by a worker, he/she will need immediate attention (washing with water, a shower, first aid or, rarely, medical attention).

- *Vapour inhalation*: wear breathable masks or respirators
- *Eye contact*: wear goggles or safety glasses
- *Skin contact*: wear suitable protective wear

6.3.3 Basic safety provisions

PU chemicals pose health hazards, so a foam operation (large or small volume) must have basic safety provisions. Safety requirements start from the initial intake of chemicals into storage to the full cycle (i.e., until the fabricated foam is sent to the holding area before being shipped).

- Protective clothing
- Provide eye-wash and shower stations
- Fire extinguishers
- Display copies of the safety datasheets of materials at strategic locations
- Provide effective exhaust systems
- Periodic in-house meetings/training on safety factors
- Occasional safety/fire-drill training by external professionals
- Regular (6 months) air-pollution checks

6.3.4 Basics of spill management

Chemical spills may not be a feature of a well-managed operation. However, they could happen and it is best to be prepared for such an event by knowing what to do. The following basic steps help with such an occurrence:

- Identify the material
- Clear the area of non-emergency personnel
- Notify the supervisor/manager
- Stop the spillage/leakage
- Contain the spilled material
- Absorb the spilled material with sawdust or industrial absorbents
- Use a solution/liquid from a spill clean-up kit to neutralise it
- After a short time, shovel the mass into one pile and possibly into a corner
- Dispose of the material mass in accordance with local regulations
- If the spill is big or uncontrollable, call for external professional help

6.3.5 Recommended safety equipment

All personnel dealing with chemicals (and even the ones in foam fabrication) need safety wear. Some of the recommended ones are:
- Protective gloves
- Protective clothing
- Respirators, masks or safety glasses
- Safety shoes
- Eyewash and water showers
- Spill-control kits
- Fire extinguishers
- First aid station
- Safety signs
- Exhaust systems

6.3.6 Handling precautions

Precautions must be observed in three major areas. The first area is unloading of the raw materials and storage in appropriate allocated areas. The second area is the weighing/metering process to the dispensing machine. The post-production stage also necessitates precautions because reactive gases are released and continue to release gases due to the exothermic (heat giving) nature of the chemical process in making foam. The post-cure area is subject to gas emission for ≈24 h and handling the foam blocks during this period is not advisable. Third, during fabrication of these fully cured foam blocks, gas is released due to heat from hotwire cutting systems (which are not allowed in most countries). Even with other standard cutting systems (e.g., band saw blades or vibrating blades) fine particles of foam dust are generated and operators must protect themselves.
- Always wear appropriate protective wear
- Never work alone if handling hazardous or reactive chemicals
- Do no inhale vapours, mists or dust
- Avoid contact with chemicals at all times
- Handle freshly polymerised foam with care
- Do not stack freshly made foam buns/blocks one on top of each other
- Equip foam-storage areas with water sprinkler systems
- Keep adequate stocks of isocyanate neutralisers to counter spills or leaks
- Never expose PUF chemicals in closed drums to elevated temperatures
- Never expose the isocyanates to water, amines or other reactive chemicals

The safety factors presented here cover a comprehensive area of a PUF operation, but depending on the size, volume of production and chemicals used, safety factors must be worked out and necessary precautions taken to ensure a trouble-free operation.

Bibliography

1. M. Thirumal, Dipak Khastigir, Nikhil K. Singha, B.S. Manjunath and Y.P. Naik, *Journal of Applied Polymer Science*, 2008, **108**, 3, 1810.
2. *Introduction to Polyurethanes: Applications*, American Chemistry Council, Washington, DC, USA, 2nd August 2015. https://polyurethane.americanchemistry.com/Introduction-to-Polyurethanes/Applications.
3. *Water-Blown Hybrid Urethane Technology*, Natural Polymers LLC, West Chicago, IL, USA.
4. *Speciality Catalysts for Water-Blown Open and Closed Cell Foams*, Air Products, Allentown, PA, USA.
5. *Blowing Agents/Foaming Agents for Foams*, Astra Polymers Compounding Co. Ltd., Al Khobar, Kingdom of Saudi Arabia. http://www.astra-polymers.com/blowing-agent.php.
6. Isopur-Bayflex Colour Systems for Foams, ISL-Chemie GmbH & Co. KG, Kürten, Germany.
7. C. Defonseka in *Practical Guide to Flexible Polyurethane Foams*, Smithers Rapra, Shawbury, Shropshire, UK, 2013.

7 Principles of foam productions, foam calculations and foam formulations

7.1 Principles of foam production

When we speak of cellular foams, there are many types of polymers that could be made into cellular materials. This book is confined to cellular foams that are blown by water instead of blown by chemicals to achieve cellular structures. Relevant terms for this subject are 'molecular weight' (MW), 'microcellular', 'foaming', 'blowing of polymer matrix', 'cell walls', 'formulation', 'polymer systems' or 'polymer blends', 'open cells', 'closed cells' and 'density'. To understand these terms and the science of producing quality cellular foams, we must understand at least the basic principles of formulation and interaction of the different components that make up a viable combination to produce good foams. In general, these are made to international standards such as Japanese Industrial Standards (JIS), British Standards (BS), American Society for Testing and Materials (ASTM) and a few others. ASTM is a good, internationally accepted standard and most manufacturers adhere to it.

First, we understand the basic chemistry of the components being used to make cellular foams.

7.2 Basic polymer chemistry

Polymer chemistry is a very complex subject. The basic information presented here is sufficient to gain good understanding of polymers and forms the basis for plastics (under which cellular plastics are classified).

The word *plastics* is derived from the Greek word *plastikos*. In general, a plastic material can be defined as a material that is pliable and capable of being shaped by temperature and pressure. Plastics are based on polymers, derived from the Greek words *poly* meaning 'many' and *meros* meaning 'basic units'. Polymers are also sometimes called 'resins' which, technically, is not correct because resins are gumlike substances.

Polymer structures, as the word suggests, are polymer materials composed of molecules of very high MW. These large molecules are, in general, referred to as 'macromolecules'. Polymers are macromolecular structures connected by covalent chemical bonds. These bonding patterns are important because they determine the physical properties of a plastic product and are generated synthetically or through natural processes. Polymers are formed when these basic units are joined together by a process called 'polymerisation'. There are two methods (addition polymerisation and condensation polymerisation) which produce synthetic polymers such as polyethylene(s) (PE), polystyrene(s) (PS), polyurethane(s) (PU), polyvinyl chloride

https://doi.org/10.1515/9783110643121-007

and many others. If a polymer consists of similar repeating units, it is called a 'homopolymer' and if two different types of units are joined by polymerisation they are called 'copolymers'. There is another class of polymers called 'terpolymers' in which three different types of basic units are joined together. Furthermore, a polymer is called 'thermoplastic' (meaning it can be re-used) and 'thermosetting' (meaning it cannot be re-used). All polymers, in general, belong to one of these two families. Examples of thermoplastic polymers are PE, PS, and polypropylenes, whereas thermosetting polymers are PU, silicones, and formaldehydes.

There are various ways that monomers (basic units) can arrange themselves during polymerisation. These can be broken down to two broad general categories: 'crosslinked' and 'uncrosslinked'. Uncrosslinked polymers can be subdivided into 'linear' and 'branched' polymers. The most common example of uncrosslinked polymers that present the various degrees of branching are PE. Another important family of uncrosslinked polymers are copolymers. Copolymers are polymeric materials with two or more monomer types in the same chain. Depending on how the different monomers are arranged in the polymer chain, they are identified as 'random', 'alternating', 'block' or 'graft' copolymers. Thermoplastics can crosslink under specific conditions (e.g. a gel formation when PE is exposed to high temperatures for prolonged periods) but thermosets and some elastomers are polymeric materials that can also crosslink. Crosslinking is the process in which some double bonds present breaks, allowing molecules to link up with the neighbouring ones. One of the main advantages of crosslinking is the material, upon solidification, becomes heat-resistant.

7.2.1 Polymerisation reactions

Polymerisation is an important stage in cellular foams, especially in cellular polyurethane foam (PUF) made with water as the sole blowing agent. In PUF production, a polyol or a blend of polyols is polymerised to produce urethane, which is then 'blown' or expanded into expanding cells and thus becomes a cellular material.

The chemical reaction in which high MW molecules are formed from monomers is called polymerisation. There are two basic types of polymerisation: i) 'chain reaction' or 'addition'; and ii) 'step reaction' or 'condensation polymerisation'. One of the most common types of polymer reactions is chain-reaction polymerisation, which is a three-step process involving two chemical entities. The first is the monomer as a link in a polymer chain. In nearly all cases, the monomers have at least one carbon – carbon double-bond. Ethylene is an example of a monomer used to make probably the most common polymer. The other chemical reactant is a catalyst (e.g., a free-radical peroxide) added in low concentrations.

The first step in a chain-reaction polymerisation is 'initiation' and occurs when the free-radical catalyst reacts with a double-bonded carbon monomer to begin a polymer chain. The carbon double-bond breaks apart, with the monomer bonding to the free-radical and the free electron being transferred to the outside carbon atom in this reaction. The next step in the process is 'propagation', a repetitive operation in which the physical chain of the polymer is formed. The double-bond of successive monomers is opened up when the monomer reacts with the reactive polymer chain. The free electron is passed down the line of the chain successively to the outside carbon atom. This reaction can occur continuously because the energy in the chemical system is lowered as the chain grows. The net result is that the single-bonds in the polymeric chain are more stable than the double-bonds of the monomer.

The third step is 'termination', which occurs when another free-radical left over from the original splitting of the organic peroxide meets the end of the growing chain. This free-radical terminates the chain by linking with the last monomer component of the polymer chain, thus forming a complete polymer. Termination can also occur if two unfinished chains bond together. This exothermic (heat giving) reaction occurs extremely fast, forming the individual chains of a polymer.

Step-reaction or condensation polymerisation is another common type of polymer formations. This method typically produces polymers of lower MW than chain reactions, and requires higher temperatures. Unlike addition polymerisation, step-reactions involve two different types of difunctional monomers or end groups that react with one another to form a chain. Condensation polymerisation also produces a small molecular byproduct: water.

7.2.2 Chemical structures of polymers

The monomers in a polymer can be arranged in several different ways. Addition and condensation polymers can be linear, branched or crosslinked. Linear polymers are made up of one long continuous chain without excess appendages or attachments. Branched polymers have a chain structure comprising one main chain of molecules with smaller molecular chains branching from it. A branched-chain structure tends to lower the degree of crystallinity and density of a polymer. Crosslinking in polymers occurs when primary valence bonds are formed between separate polymer chain molecules. Chains with only one type of monomer are known as 'homopolymers'. If two or more different types of monomers are involved, the resulting copolymers can have several configurations, such as random copolymers, alternating copolymers, block copolymers and graft copolymers. If three different types of monomers are used, then terpolymers are formed.

7.2.3 Physical structures of polymers

Segments of polymer molecules can exist in two distinct physical structures: 'crystalline' or 'amorphous'. Crystalline polymers are possible only if there is a regular chemical structure (e.g., homopolymers or alternating copolymers) and the chains possess a highly ordered arrangement of their segments. Crystallinity in polymers is favoured in symmetrical polymer chains, but it is never 100%. These semi-crystalline polymers possess a rather typical liquefying pathway, retaining their solid state until they reach their melting point.

Amorphous polymers do no show order. The molecular segments in amorphous polymers or the amorphous domains of semi-crystalline polymers are arranged randomly and entangled. Amorphous polymers are at low temperatures below their glass transition temperature (T_g), the segments are immobile and are often brittle. As temperatures increase close to T_g, the molecular segments can begin to move. Above the T_g, the mobility is sufficient to enable the polymer to flow as a highly viscous liquid. The viscosity decreases with increasing temperatures and decreasing MW. There can also be an elastic response if the entanglements cannot align at the rate a force is applied, and then this material has a viscoelastic effect.

7.3 Basic components for water-blown cellular foams

Ninety percent of water-blown cellular foams are based on flexible or rigid cellular PUF, so this discussion is based on them. The leading chemicals manufacturer BASF offers a full range of components: polyether polyols, polyester polyols, diphenyl methane diisocyanate(s) (MDI), toluene diisocyanate(s) (TDI), tertiary amines, and surfactants.

- MDI is a versatile isocyanate that can be used to make flexible, semi-rigid and rigid cellular foams. Its primary applications are insulation, furniture, interiors, automotive components, and shoe soles.
- TDI is an isocyanate used primarily in the manufacture of flexible cellular foams. Its primary applications are for cushions, mattresses, automobile seats and other products in comfort applications.
- Polyether polyols form the basic ingredient for making flexible and rigid foams.
- Polyester polyols (aliphatic and aromatic grades) are available. They are combined with isocyanates to produce semi-rigid foams.
- Amines (generally tertiary amines) are used and act as catalysts.
- Surfactants: usually, the best results are obtained with silicone surfactants, which control the cell-formation phase.
- Tin catalysts (generally stannous octoate) are used to catalyse the polymerisation reaction and prevent splits in the foam mass.

- Fillers are used to reduce the costs of foam as well as to increase density and mechanical properties.

PU are the reaction product of an isocyanate and a hydroxyl group. 'Poly' denotes that more than one urethane group is involved. Typically, diols or polyols ('di' refers to two functional groups and poly refers to more than two functional groups) and a diisocyanate or polyisocyanurate is used to produce a PU. A hydroxyl group is a reactive group which consists of one hydrogen and one oxygen. Water, for example, consists of two hydrogen atoms and one oxygen atom. It is also a hydroxyl functional compound and, therefore, reacts with an isocyanate. What makes the PU so interesting is the very fast reaction of a hydroxyl group with an isocyanate. Under appropriate conditions this reaction is completed in seconds. A second aspect which is of great importance is the high stability of the isocyanate linkage to water under acidic and basic pH conditions. The urethane group contributes excellent properties to a polymer. Depending on the composition of the PU, the polymer can have excellent water resistance, toughness, abrasion resistance, light stability and chemical resistance. For most PU formulations, close control of the ratio of the isocyanate to hydroxyl component is required.

In the presence of water, the isocyanate reacts with water. This leads to formation of carbon dioxide (CO_2) and gassing or foaming occurs. The amine catalyst promotes formation of a urea linkage very rapidly. The reaction of an isocyanate group with water is used in one-component cure coatings, adhesives or sealants as a cure mechanism. Pre-polymers with isocyanate end groups are prepared. These polymers must be stored under dry conditions and cure in the presence of moisture or water vapour. PU can also be stable to ultraviolet (UV) light and outdoor exposure if aliphatic isocyanates are used to make a polymer. PU prepared from aromatic isocyanates are very sensitive to UV degradation and must be covered or coated.

7.3.1 Diphenyl methane diisocyanate-based cellular foams

Obtaining low-density (<30 kg/m³) in all water-blown MDI-based foams without compromising mechanical performance (tensile strength and tear strength) and compression has been a significant challenge for PUF manufacturers. It is straightforward to make such foams with TDI, but with MDI it becomes very difficult.

With some properties, such as flame retardance, ability to obtain harder grades without use of copolymer polyols, and faster cure times, MDI foams offer a distinct advantage over TDI-based systems. The real challenge with MDI-based systems is to obtain durable foams at low densities in all water-blown formulations. One of the important reasons why it is so difficult to create MDI-based foams of low density is the large increase in the amount of water required to get low densities in such foams. After about four parts of water per hundred parts of polyol in the formulation, it becomes increasingly difficult to reduce density by further increases in water content. Furthermore, the amount of hard segments increases approximately linearly with the amount of water. This causes the urea content being formed to rise rapidly with very little gain in reduction of density. Hence, it becomes a case of diminishing returns if high amounts of water are used.

Experiments have shown that, to obtain a 20 kg/m^3 MDI-based foam, ≈5 parts of water per hundred parts of polyol must be used compared with only ≈4 parts per hundred of polyol for TDI-based foams. The MW per molecule of urea for a MDI-based foam is higher than for a TDI-based urea due to the higher MW of a MDI molecule compared with TDI. Another set of challenges that MDI-based formulations have over TDI-based systems is due to the presence of higher oligomers (polymers) that can cause poor hard-phase organisation in MDI-based systems. The higher oligomers also increase the functionality of the isocyanate to >2. Higher functionality of the isocyanate can cause faster reactivity and gelling, thereby preventing good processing characteristics.

7.3.2 Toluene diisocyanate-based cellular foams

The most common and easy way to produce cellular PUF are TDI-based cellular foams with water as the sole blowing agent. Polyether polyols, polyol blends and graft copolymers or a suitable polyol system can be used. Foam systems are generally formulated with the main polyol as a base and taken as 100 parts by weight (pbw). All other components used in the formula are calculated by weight or percentage of this base number. In general, amines (catalysts), silicones (cell controllers) and calcium carbonate (filler), tin catalysts, together with water as the blowing agent produce good basic cellular foams. Water content plays a key part in producing CO_2, which turns the polyol polymer into cellular material.

Density is a key factor in cellular foams, ranging from very soft to medium to heavy. Blowing agents must be used within certain limits to control foaming and prevent fire from the already exothermic (heat giving) reactions taking place. Increasing the blowing-agent and TDI content results in lower densities and lower contents give higher densities. Limits of water usage are 2.0–5.0 pbw of the total polyol content in a formulation but some producers may use ≤6.0 pbw. Taking a general range of densities of 16–32 kg/m^3, then 2–5 pbw is fine or ≤6 pbw. Higher densities can be achieved by using fillers and other additives, but to achieve densities <16.0 kg/m^3, another auxiliary blowing agent (e.g., methylene chloride) must be used.

To achieve any end-properties desired and aesthetic values, many additives and colourants are available. Biomass ash can be used as part of the filler content. Of particular interest is the very fine 'fly ash' or very fine rice hulls ash. The high content of silica in the latter case greatly helps to enhance the mechanical properties of cellular foams. Rice hulls ash already comprises lignin (a polymer)

and has ≈20% silica, and burnt ash contains ≈70–80% silica. There are companies (especially in the East) who supply these items to customer specifications.

7.3.3 Non-isocyanate polyurethane foam

An international nanotechnology research centre called Polymate Limited has been
developing non-isocyanate curable polymer systems and foam materials based on
hybrid non-isocyanate polyurethane foam (NIPF). Thus, a new type of specialty-
amine derivative has been developed and is being used as curing agents for epoxy
compositions. NIPF combines the best properties of epoxy and PU systems, but
does not have the major inherent disadvantage of conventional PUF, high toxicity
of isocyanates, because they are not used.

It has been reported that, in NIPF production, ≤50% of raw materials are from
renewable materials such as vegetable oils. The uniqueness of the structure and
properties of these new hybrid systems is that they allow foaming compositions in a
wide range of physical properties: flexible, semi-rigid, rigid and soft. NIPF also has
additional features that extend the range of applications to high fire resistance, and
chemical resistance. Applications of these foams can use standard methods such as
spraying, pouring, injection and foam moulding, as both physical or chemical blowing agents can be used to produce the gas-phase. These systems are two-component
systems and cellular foams can be produced in a factory using conventional methods
or on site. On the basis of hybrid non-isocyanate compositions, Polymate Limited reports that they have foam formulations for insulation, packaging, construction, furniture, and special-purpose foam. The only constraint to overcome could be the lower
yield of foam volume in non-petroleum-based polyols as compared with petroleum-
based conventional polyols.

7.4 Spray polyurethane foam insulation

Self-curing water-blown spray polyurethane foams (SPF) can be an alternative to
conventional insulation such as fibreglass. This specialty spray foam is created by
combining a two-component mixture of an isocyanate and a soya-based PU resin.
The two components come together at the end of a spray gun and form an expanding mixture that can be sprayed onto almost any surface to produce very efficient
insulation. Many residential, commercial and industrial builders are making spray
foam insulation their first choice for insulation.

Some of the benefits of spray foam insulation are:
- Reduction of energy consumption: according to research, ≈40% of the energy
  (inside heat) of most buildings is lost if insulation is absent. Spray foams eliminate this heat loss almost completely, especially around crevices, nooks,

windows, doors and hard-to-reach places because the liquid foam can flow and expand, thus providing tight insulation.

– Thermal effectiveness: spray foam insulation has a higher insulation value or resistivity factor (R-value) with regard to comfort. If combined with characteristics such as air tightness and efficient sealing, it enables heating, ventilation and air-conditioning to work at lower levels but with identical efficiency, thus reducing costs tremendously. Alternatively, smaller air-conditioning and heating units can be used.

– Air tight sealant: it is easy to blow through traditional fibreglass insulation but not so in spray foam insulation, where a light top-skin also forms after full expansion of the foam. If the foam liquid is applied to a surface through a pressurised nozzle of a spray gun, the liquid 'grips' the surface first before expanding. Thus, a combination of very low air permeability and continuous sealed application ensures much better insulation for a building.

– Vapour permeance: closed-cell spray foams of ≈1 kg are sufficient to provide vapour diffusion in keeping with international building codes.

– Durability: tests have shown that these PU sprayed insulation foams have very high tensile and compressive strengths that are better than those for fibreglass. They also add an element of structural integrity to a building. Also, these unique spray foams are not susceptible to water damage and last the lifetime of a building.

– Noise reduction: spray foam insulation results in a noticeably reduced amount of noise and can be very beneficial in places where excessive noise could be generated (e.g., laundry room, music rooms, living rooms).

– Air quality: with the building sealed tighter, less noise pollution and water can leak into the building. Hence, air filtered through intake/outtake points and walls is less susceptible to mould and mildew growth.

Some of the most common and popular areas of spray-foam application are:
– Exterior walls: vapour barrier, thermal insulator, auditory insulator.
– Concrete living spaces: vapour barrier, thermal insulator.
– Patio living spaces: vapour barrier and thermal insulator.
– Insulating pools: thermal insulation.
– Below basement slabs: thermal insulator.
– Damp-proof foundation: vapour barrier and thermal insulation.
– Industrial/commercial/freezers: vapour barrier and thermal insulation.

7.5 Calculations for water-blown foams

Foam calculations are complex, especially because chemicals are involved. Cellular foams have a wide range of applications with different requirements for the many end-applications possible. Hence, a foam producer must have thorough knowledge of the function or functions of each chemical ingredient to be used in a formula, especially due to the steady flow of new products/ingredients coming onto the market through research and development. These new products have advantages over conventional ones. Also, due to the environmental concerns of past and current applications, foam producers must deal with eco-friendly components because the foam industry is slowly moving away from traditional petroleum-based products. A good example is the transition of petroleum-based polyols to bio-based polyols.

Also, formulations must also be geared to the dispensing methods. Water is the sole blowing agent, so knowing the exact relationship of water to the isocyanate and polyol/polyols in a formulation is important. The key is calculation of the required isocyanate content in relationship with polyol/polyols, which is taken as 100 pbw as the basis of a formula.

A recipe for achieving a desired foam may come from previous work and experience, or may be unique in that it is based on emerging new technologies. The first step is to list all the ingredients or components which are used. A foam producer knows by experience or through assistance the exact specifications of all components to be used and their functions in relation to each other.

The amount of isocyanate required to react with the polyol, water and other reactive additives to give the desired stoichiometry is shown below. Many foams are prepared at a slight excess of isocyanate (e.g., 105% of theoretical equivalence). In practice, the amount of isocyanate is adjusted upwards or downwards depending on the particular foam system, the properties desired, the ambient conditions and the scale of manufacture (especially for continuous foaming systems).

In foam productions the basic units used are:
- Mass (weight) = volume × density
- Mass = g or kg
- Volume = cm^3 or m^3
- Then density = g/cm^3 or kg/m^3

7.5.1 Mixing ratio

For a small-to-medium producer of foam, the easiest way is to use a two-component system: component A (isocyanate) and component B (polyol + other ingredients). However, only one density can be produced per system whereas, if a foam producer was to buy all ingredients separately, different densities can be made by combining the ingredients in different combinations.

A mixing ratio calculates the specific amounts of isocyanate and polyol blend needed for reactions to produce good foams. In general, a raw-material supplier indicates a mixing ratio but if a foam producer needs to formulate a two-component on his/her own, the following information given in the certificate of analysis by the supplier is required:

- The hydroxyl number (OH#) of the polyol
- The amount of water as a percentage present in the polyol
- The isocyanate content NCO group

$$\text{For example}: \quad \text{MDI} = \frac{(\text{polyol OH} \times 100) + (\% \text{ water} \times 6{,}233)}{(\text{NCO content} \times 13.35)}$$

This is the amount of MDI to add to every 100 parts of polyol.

Example: A polyol blend has an OH# of 95 and water content of 0.45%. The MDI has a NCO content of 23%. What is the mixing ratio?

$$\text{MDI} = (95 \times 100) + (0.45 \times 6233)/23 \times 13.35$$
$$= 12{,}305/307 = 40.1$$

Therefore, 40.1 parts of MDI are needed for every 100 parts of polyol.

As Table 7.1 indicates, all recipes and calculations are based on a total of 100 pbw of polyol. There may be more than one polyol in a formulation but the sum of these must add up to 100 parts. The amount of other ingredients is calculated and listed using 100 polyol as the basis.

Table 7.1: Examples of formulations.

Formulation	Quantity (pbw)	EW	Equivalents
Polyol	70.00	1,825.00	0.0383
Copolymer polyol	30.00	2,400.00	0.0125
Surfactant	1.00	0.00	–
Pure amine	1.70	35.00	0.0485
Catalyst 1 (Dabco)	0.15	105.00	0.0014
Catalyst 2 (Niax)	0.08	233.70	0.0003
Catalyst 3 (Polycat)	0.24	0.00	–
Water	4.20	9.00	0.4666

Formula weight = 107.370
Total EW = 0.5676
Isocyanate T-80 TDI EW = 87.1

Index	Isocyanate requirement	Components ratio
90.00	44.51	2.4123
95.00	46.99	2.2850
98.00	48.47	2.2152
100.00	49.46	2.1708
103.00	50.94	2.1078
105.00	51.93	2.0676
108.00	53.42	2.0099
110.00	54.41	1.9734

EW: Equivalent weight

The calculation for a foam formulation is straightforward:

- Determine the parts of each polyol. Total should be equal to 100 (A).
- Determine the parts of other B-side components per 100 parts of polyol.
- Sum the parts of all B-side components to get the total formula weight.
- Record the EW of each B-side component from the calculation shown above or from listing of typical EW as shown in the tables.
- Calculate the equivalents of each B-side component.

Contents of water in standard polyols are very low and can be ignored when calculating the actual total water and no adjustment is needed. However, some copolymer polyols have an appreciable content of water and, if very high densities are formulated, then an adjustment for this water content may be necessary.

7.5.2 Ratio calculation

In foam formulations, the three main components that determine the final density are polyol, isocyanate and water. In a two-component system, taking polyol (+ water + others) as the A component and isocyanate as component B, it is important to calculate the correct ratio between them.

For example: Formula ratio of 100 parts of component A to 160 parts component B pbw.

Taking the actual metered weight of A as 1,200 g/min, then determine the throughput of B to correspond to the given ratio then:

$$\text{g/min Part A} \div \text{g/min Part B} = 100 \div 160 = 1{,}200 \div \text{Part B}$$

Therefore:

$$100\ B = 1{,}200 \times 160$$
$$B\ \text{throughput} = 1{,}920\ \text{g/min}$$

Contents of water in standard polyols are very low and can be ignored when calculating the actual total water and no adjustment is needed. However, some copolymer polyols have an appreciable content of water and, if very high densities are formulated, then an adjustment for this water content may be necessary.

7.5.3 Calculation of density

In foams, probably the most important first property is density. The two main components that influence the density of a foam are isocyanate and water. The higher the contents of these two, the lower are the densities and *vice versa*. There are limits for both components with a maximum of 6.0 pbw for 100 polyol. In a general range of density of 16–32 kg/m³, water can be used as a single blowing agent, whereas higher densities can be achieved by using graft copolymers and other conventional fillers such as calcium carbonate. I would like to introduce the use of biomass ash as part of the filler content. For densities <16 kg/m³ (very light), an auxiliary blowing agent such as methylene chloride can be used but, in the formulations projected later, auxiliary blowing agents are not considered because this book is based on water-blown foams.

Density can be defined as mass (weight) per unit volume. The formula for the calculation is:

$$M = V \times D$$

Where:

M: mass in g or kg;
V: volume in cm³ or m³; and
D: density in g/cm³ or kg/m³.

7.5.4 Calculation of indentation force deflection

Next to density, the indentation force deflection (IFD) is the second most important property in foams. In general, the marketplace would look for density and IFD as the two primary factors for quality. The IFD is an indication of the ratio of compression of a selected foam. This is especially important in the bedding and furniture industry. If the IFD is ≥2.0, the foam is acceptable, whereas foams having IFD <2.0 are classified below par.

The IFD factor is the ratio of compression of a selected foam sample of size 60 × 60 × 10 cm being compressed to 25 and 65% using an indenter plate on an electromechanical device. The ratio, calculated as the 65% reading and 25% reading, is known as the IFD factor.

7.5.5 Calculation of material for a given formula

Correct formulation is essential for the production of quality foams. Many formulations are available for different sources or one may develop these through experience or by trial and error. Table 7.2 provides recommendations for a practical and workable formulation and can be used as an example (the total foam batch required is = 98.2 kg).

7.6 Foam formulations

The formulations shown in Table 7.3 are fairly accurate, but they are only recommendations and should be used as guidelines.

Table 7.2: Calculations for raw-material components.

Component	pbw	Quantity	Weight (kg)
Polyol	100.00	100.00/175.95 × 98.2	35.81
TDI	70.00	70.00/175.95 × 98.2	39.07
Water	4.50	4.50/175.95 × 98.2	2.51
Surfactant	0.60	0.60/175.95 × 98.2	0.33
Amine catalyst	0.30	0.30/175.95 × 98.2	0.17
Filler	0.35	0.35175.95 × 98.2	0.20
Tin catalyst	0.20	0.20/175.95 × 98.2	0.11
Colour	–	–	–
Additives	–	–	–
Total	175.95	–	98.20

Table 7.3: Typical formulations for basic foam products.

Component	Standard (pbw)	Mattresses (pbw)	Pillows (pbw)
Polyol	85.00	80.00	100.00
Graft	15.00	20.00	–
TDI	40.70	70.00	84.00
Water	2.90	3.50	4.50
Surfactant	1.20	0.60	0.30
Amine catalyst	0.12	0.70	0.27
Tin catalyst	0.04	0.03	–
Flame retardant	3.00	–	–
Foam properties			
Density (kg/m3)	27.2	20.8	17.6
Support factor (IFD)	2.30	2.10	2.00

7.7 All-water-blown rigid foam system

Ecofil 40 is a two-component all-water-blown rigid PUF system. Typical moulded densities are ≈40–50 kg/m³. This system produces a fine cell foam with low friability and good adhesion to substrates. Resin component (B) is a fully formulated liquid mixture and should be stored in tanks or drums and heated to 20 °C before use. Approximate viscosity at 20 °C = 1,000 cps and specific gravity at 20 °C = 1.11.

Component (A) is also a liquid mixture containing MDI and also should be stored in sealed tanks or drums and heated to 20 °C before use. Viscosity at 20 °C = 325 cps and specific gravity at 20 °C = 1.23.

Processing equipment should be suitable dispensing machinery to produce foam from processes such as free rise, pour-in-place or moulding. These machines should be capable of maintaining a mix ratio of ±2% accuracy and temperature at 20–25 °C (Table 7.4).

Table 7.4: Typical physical properties of Ecofil 40.

Property	Value	Standard
Core density	38 kg/m^3	BS 4370
Compressive strength	Parallel to rise: 240 kN/m2	BS 4370
	Perpendicular to rise: 180 kN/m2	BS 4370
Tensile strength	200 kN/m^2	–
Buoyancy	958 kg/m^3	–
Closed-cell content	94%	–
Thermal conductivity	At 23 °C: 0.030 W/m °C	ASTM C-518

7.8 Calculations for water-blown large foam blocks

Take the case of water-blown large foam blocks poured manually in a semi-automatic operation or in a more sophisticated dispensing machine. In the production cycles, large foam blocks are made one by one, with the advantage of making different densities/properties with each pour. A foam producer must ensure that sufficient material in a batch is available to cover the whole mould up to the calculated full-rise time. If there is too much material, it overflows over the mould and is wasted. When calculating the exact quantities for a particular size of foam block, a gas evaporation factor of 0.5–1.0% and provision for the thickness of the skins are needed. An ideal procedure would be to formulate and test a very small batch using a box 12 × 12 × 6 inches (box test), which provides a lot of data and shows if the formulation needs modification.

Example: To make a foam block of dimension 80 × 60 × 28 inch

$$M = V \times D$$

Where:

M: mass;

V: volume; and

D: density 4.0 lb/ft^3.

M = [(81.5 × 61.5 × 28) inch^3 × 4.0 lb/ft^3] ÷ 1,728 inch^3

= 324.86 lb = 147.7 kg + 1% = 149.2 kg

Therefore, a single pour or a dispensing machine must have a single-shot capacity of ≥150 kg or the foam producer could make smaller blocks. Depending on the operation, a small chemical-wastage factor may also be considered.

Now, consider a two-component raw-material system for making memory foam blocks:

- Mixing ratio component B (polyol blend) and component A (isocyanate) = 100:31;
- Cream time: 11 s; and
- Rise time: 143 s.

Component B = 100 ÷ 131 × 149.2 = 113.89 kg
Component A = 31 ÷ 131 × 149.2 = 35.31 kg

Bibliography

1. *Dow Polyurethanes – Foam Preparation Calculations*, Dow, Midland, MI, USA, 2014. http://dowac.custhelp.com/app/answers/detail/a_id/5731/~/dow-polyurethanes-foam-preparation-calculations.
2. *All Water Blown Foam System*, Version 2, Isothane Ltd, Accrington, UK, 2006.
3. http://www.wernerblank.com/polyur/learning/learning_center1.htm.
4. *Benefits of Polyurethane Spray Foam Insulation*, Coast Spray Foam Ltd., Abbotsford, Canada.
5. C. Defonseka in *Practical Guidelines to Flexible Polyurethane Foams*, Smithers Rapra, Shawbury, Shropshire, UK, 2013.

8 Practical processing methods of water-blown cellular polymers

In this chapter, production methods and procedures for water-blown cellular polymers are presented in detail for the benefit of foam manufacturers and especially for entrepreneurs and small-to-medium foam producers. I discuss challenges that producers must face and recommend solutions with cost reductions in mind. Where organic fuels are used currently, interesting inorganic alternate fuels are recommended as tried and tested for efficiency and with considerable cost savings. Readers may find exciting new possibilities of using biomasses based on emerging technologies as part of traditional fillers and stiffening agents.

8.1 Self-skinning moulded flexible foam

S-Foam 48 is a two-component water-blown flexible polyurethane foam (PUF) system used especially for low-density moulding. Nominal density of the finished product is 40–56 kg/m^3 if open blown or not restricted. Higher densities may be achieved in closed or restricted moulds.

8.1.1 Storage of raw material

Store both containers in an area where the temperature is 21–32 °C. When using the material, check a small sample visually to make sure that no crystallisation is present. Crystallisation can occur during shipment and in cold weather. If the contents are cloudy or gummy, both components should be warmed with the containers open and stirred until the material returns to its appropriate smooth liquid consistency.

8.1.2 Mould preparation

The moulds should be well waxed or sufficient release agents should be applied for easy release and sealed securely. Foams may seek moisture through release waxes and stick to mould surfaces if insufficient release is not available. The type of release agent to be used depends on the foam material. The mould should be warmed to 24–29 °C before filling the mould. Once the mould is heated and the production cycle has started, it stays warm for continued moulding cycles. Release systems vary in accordance with mould materials but, as a general rule, the manufacturers of this two-component foam system recommend JWax, Challenge 90 and Cearra

https://doi.org/10.1515/9783110643121-008

Wax. Silicone waxes are not suitable because they cause poor surface finish and prevent the adhesion of paints and over-coatings, where applicable.

The ideal moulds for long production runs are machined aluminium or epoxy moulds. Epoxy moulds offer the least expensive method for long-term use, when cycle times allow slower dissipation of heat (Table 8.1).

Table 8.1: Processing parameters.

Mixing ratio	Part A		50 pbw
	Part B		100 pbw
Viscosity	Part A	cps at 25 °C	205
	Part B		550
Cream time	At 25 °C		25 s
Rise time	At 25 °C		2.5–3.5 min
Demould time	At 25 °C		15–30 min

8.1.3 Processing method

Weighing materials separately rather than pouring together on a scale is the preferred method because it allows for more time when combining the materials and prevents premature reactions. Both components should be weighed carefully in accordance with supplier recommendations to achieve the correct mixing ratio. As a general rule, both components should be pre-warmed at 24–29 °C. Colder temperatures can cause sluggish and poor expansion of the foams. Excessive heat causes the foams to react quickly and may cause poor cell structure or cause the foam to collapse.

Mixing can be done with a high-speed electric drill or an air motor with a suitable mixer blade which gives high shear and thorough mixing within 5–8 s (which is sufficient in this case). The material should have a uniform blended appearance before being poured into a mould. Mixing too long or not enough can result in poor material performance. The ideal solution to this problem is to carry out a 'box test': try out a small batch in a wooden box of dimension 12 × 12 × 6 inches. The box test indicates the correct cream time, rise time and cure time and, if it is cut in the middle, the cell structure and density can be measured. Importantly, once mixed, the material should be poured quickly into the mould. If done too slowly, the foam rises in the mixing container and the batch is lost. Hence, the mixed material must be poured into the mould in the liquid state before it 'creams' (foams).

8.2 JHMorrow open-cell spray polyurethane foam

JHMorrow open-cell spray polyurethane foam (SPF) insulation material is an open-cell, two-component, low-density, fully-water-blown system designed specifically for insulation applications. This technologically advanced and economical insulation system is ideal for insulation of buildings, providing improved comfort for occupants, a cleaner indoor environment, greater reduction in noise, and superior energy savings compared with conventional insulation systems.

8.2.1 Unique properties

This low-density SPF insulation expands 120:1 from its original liquid state to fill cracks, voids, crevices and building cavities to provide a climate-controlled building environment. By mitigating airflow through and within walls (incoming and outgoing), heat and cold transfer (thermal conductivity) moisture accumulation in the building materials reduce the chance for mould and mildew to accumulate and grow. These foams also minimise the transfer of sound. This insulation foam does not settle, shrink or deteriorate and is guaranteed to retain its R-value and noise-reduction properties over the lifetime of the building.

8.2.2 Processing method

The raw material is self-curing and can be applied using conventional spray guns with adjustable nozzles. It is advisable for the operator to wear adequate protective clothing. When purchasing this insulation material, builders should adhere strictly to the dispensing conditions and correct proportions to ensure the right mixing ratio. The thickness of the spray-foam layer is determined by the specifications of insulation required for each building.

8.2.3 Recommended uses

This insulation foam can be used for residential, commercial or industrial structures. It can also be applied to the underside of roof decks or in non-vented crawl spaces for greater energy savings. Some of advantages are:
- 'Green' insulation material developed from natural resources
- Uses an environmentally friendly and non-ozone-depleting blowing agent
- Outperforms fibreglass insulation by 3–4-fold
- Makes an airtight seal by expanding *in situ*
- Foam does not affect the environment and is economical
- Foam allows insulation of difficult-to-reach places
- Foam is a partial airflow barrier, reducing dust and pollen
- Foam minimises the transfer of sound

This special foam has been developed by JHMorrow Spray Foam Insulation Company.

8.3 Two-component system for pour-in applications

SEALECTION® PIP is a two-component open-cell, semi-rigid PUF system made by Demilec Company. It is formulated specially for pour-in-place applications. This system is fully water-blown and can be used as a thermal insulation for filling wall cavities. This system complies fully with the residential and commercial building codes for SPF plastic insulation set by the International Code Council (Table 8.2).

Table 8.2: Profiles for liquid components.

Property	Component A (polymeric MDI isocyanate)	Component B (resin)
Colour	Brown	Amber
Viscosity at 25 °C	180–220 cps	250–450 cps
Specific gravity	1.24	1.09–1.10
Shelf-life	12 months	6 months
Storage temperature	10–37.8 °C	10–37.8 °C
Mixing ratio	1.1	1.1

Reactivity profile			
Mixing method	**Cream time**	**Gel time**	**Tack-free time**
Hand mix*	13–16 s	46–53 s	66–74 s
Machine mix	3–5 s	15–20 s	20–25 s

* Hand mix: mixing speed at 2,500 rpm for 10 s and liquid components at 20 °C.
MDI: Diphenyl methane diisocyanate

Temperatures and pressures for foam applications can vary depending on site temperature, humidity, elevation, substrate, equipment and other factors. During processing, the applicator must observe the characteristics of the sprayed foam and adjust processing temperatures and pressures to maintain appropriate cell structure, adhesion, cohesion and general foam quality.

Equipment used must be capable of delivering the proper ratio (1:1 by volume) of polymeric (MDI) and polyol blend at adequate temperatures and pressures. Substrate must be ≥5° above the dew point, with best processing results obtained when ambient humidity is <80%. Substrate must be free of moisture (dew or frost), grease, oil, solvents and other materials that would affect adhesion of the foam adversely. This product should not be used if the continuous service temperature of the substrate or foam is below −51.1 or >82.2 °C. SEALECTION® PIP should not also be used in contact with bulk water or to cover flexible ductwork.

8.4 Processing flexible polyurethane foam with non-traditional additives

Based on my research and in order to meet exciting new emerging needs of the flexible PUF market, I present the recommendations shown below.

8.4.1 Alternate fillers for polyurethanes

Calcium carbonate is an inorganic mineral produced from natural sources and through synthesis. Calcium carbonate is among the most popular and cost-effective fillers among others used by PUF manufacturers. The three main functions of calcium carbonate are to increase density, increase compression strength, and cost reduction. An alternative is using biomass ash, especially wheat husk or rice husk ashes, which contains 70–80% silica. Rice husk ashes also provide increases in density and compression strength, cost reduction, moisture resistance and an environmentally friendly product, and can be used as part of conventional fillers.

8.4.2 Improving thermal conductivity of flexible foams

Many advances are taking place in the automobile industry, so some parts suppliers are looking to improve the thermal conductivity of flexible PUF, particularly for cushions and seating. Realisation of the potential of graphene (a mineral found in the earth) having super electrical power conductivity and strength has scientists and researchers excited. This mineral in a very fine powder used in a polyurethane (PU) formula ≈4–6 pbw increases the thermal conductivity of that material. To ensure good compatibility, it can be mixed with the fillers being used or dissolved in the polyol blend. For improving the thermal conductivity of water-blown rigid PUF, expandable polystyrene (EPS) beads can be used. The gas generated by the exothermic reaction expands the EPS beads and blends with the PUF being generated by the polyol, isocyanate and water reactions.

8.4.3 Simple method for producing flexible polyurethane foams

A simple method for making water-blown flexible PUF is discussed here. Products that can be made are cushions, slabs, contoured shapes, and pads. They have applications in the furniture industry and suit an entrepreneur with limited resources and space. This method can be classified as the 'intermittent method' whereby quality flexible foams can be made, though the production volumes are small (≈2–3 tonnes per month).

Furniture manufacturers may need cushions in many sizes but two of the basic popular ones are 50 × 50 × 10 cm and 50 × 50 × 6.25 cm. The box mould can be made of 1-inch thick wood with grooves to fit into one another to give a tight fit and prevent leaks. All four sides must be removable and detachable from the base for demoulding of the foamed block. The dimensions of the wooden mould depend on the sizes of foam blocks to be made. Provision must be made for 1.25-cm skin waste on all sides (which also includes a shrinkage factor).

For mixing, a single-phase electric drill (with forward/reverse) having a speed range 900–1,200 rpm would suffice. This drill has a steel shaft attachment (length 75 cm and diameter 1.25 cm) and a round flat disc (diameter, 10 cm) at the bottom. For better mixing, this disc could have vertical flanges placed strategically. The design of this disc should be to achieve good homogeneous mixing in seconds.

Other basic equipment is a small weighing machine, protective wear, a floating lid constructed out of very light material (plywood), a few plastic buckets and a few small containers for weighing the separate ingredients. For this setup, laboratory test tubes could be used effectively. If hotwire cutting is allowed, then an entrepreneur can fabricate one or use a band-saw cutting machine. A foam producer has the option of using a two-component raw-material system component A (isocyanate) and component B (polyol blend). The systems supplier provides the mixing ratios and process guidelines, and the production is simple. If the rising foam reaches about one-third of the mould height, it starts to form a meniscus (dome) and the floating lid should be applied just before to prevent a rounded top (which is a waste).

If individual components are used, weigh and add any colour desired to the polyol and mix for 60 s. Mixing in air from an outside source to create more 'bubbles', is an option. If fillers and additives are used, add them to the polyol. Weigh and mix the water, surfactant and tin catalyst in a glass test tube and the mixture turns milky. Add this to the polyol mix and mix again for 8 s. Add the amine catalyst and mix again for 6 s. By now the toluene diisocyanate (TDI) (isocyanate) should have been weighed in a separate bucket and kept ready. Add the TDI into the polyol blend mix and mix quickly for 4 s and pour it into the ready mould in the liquid state. If there is a delay, the mix starts to 'cream' inside the bucket and the batch is wasted.

The liquid at the bottom of the mould lined with a thin plastic sheet creams and starts to rise slowly. When it has reached about one-third the height of the mould, quickly place the floating lid on top of the rising foam. When it has reached full height, remove the lid and allow to cool/cure. Test the surface and, when it is not tacky, the foam block can be demoulded and sent to the post-curing area and stored with 1-foot distances separating them. This precaution is because an exothermic (heat giving) reaction is taking place and fire hazards must be prevented. After 24 h, these blocks can be cut and fabricated as desired.

8.5 Moulding water-blown expandable polystyrene

Here, I present a case with hands-on experience that could be an ideal project for an entrepreneur or small producer of EPS products. This is a low-cost investment project with good profit margins and based on a semi-automatic operation.

- *Raw material* is water-blown expandable polystyrene (WEPS) in the form of tiny spherical beads consisting of two sizes of beads: small for the moulded products and slightly larger beads for the large blocks (from which sheets are cut). For ease of handling and movement, these are 25-kg paper bags or 220 kg per steel drum. If coloured products are needed, self-coloured beads must be purchased (more expensive) because colouring during processing gives a surface colour only.
- *Products*:
  - Large blocks cut into different thicknesses for building insulation
  - Standard fish boxes for local transport and export of fish
  - Fishing floats for fishing nets used by deep-sea fishing boats
  - Hot/cold container packs for food and ice cream
  - Insulating pipes for hot/cold applications

  Approximately >2 metric tonnes of raw material can be processed per month.
- *Energy source* is a steam boiler with specially designed feed arrangement for delivery of solid fuel. Approximate specifications are 50 psi, 150 kg steam/h, preferably with water-treatment steam accumulator, safety valve, and water gauge. These arrangements allow easy access to and removal of the ash residue at the bottom. They also have a downward duct on the chimney to bring the very fine ash dust that rises ($\approx$8%).
- *Fuel*: instead of using conventional diesel oil or coal, solid biomass fuels such as wheat husk and rice husks in the form of pellets or wood chip pellets can be used. These are environmentally friendly and reduce the volume of air pollutants considerably. Moreover, the resulting residues have a ready market and can be converted readily into valued-added products.
- *Machinery*: pre-expander WEPS cannot be pre-expanded on standard convention pre-expanders. They must be modified or specially designed ones must be used because WEPS, being water blown, needs extra energy (heat) to 'blow' (unlike the standard pentane gas-blown EPS). Part-expanded beads have lumps that can be neutralised by a vibrating fluidised bed (optional) or, if labour is available, broken up manually, to ensure free-flow beads. This material can be stored in a silo or in large cotton cloth bags. Unlike pentane-blown EPS, this material does not need further 'maturing' and can be moulded immediately.
- *Moulds* consist of one large block mould of size 1 × 1 × 0.5 m designed to allow even distribution of steam from all sides and also have water inlets to cool the

moulded block later. Ideally, all the sides of this block mould have perforations as steam inlets and also have two or three sides opened to remove the cooled and moulded block.

The other moulds should be made of aluminium in two halves and have perforations as steam inlets. These can be mounted on individual, small, hydraulically operated platens, with one-half of the mould fixed to the base platen and the other half movable upwards and downwards. Moulded products can be ejected by air or manually. Here, depending on costs, the moulded products can be vacuum-cooled or cooled by water.

- *Cutting machine*: a single-wire or multiple wire hotwire cutting machine would suffice to slit the large EPS blocks into sheets. This machine can be fabricated easily to save costs. The basic setup is a wooden horizontal bed (motorised) or a tilted table at 45° (gravity feed), with two aluminium channel arms fixed vertically. A hotwire (nickel/chrome) wire is fixed across these two arms, with provision for movement upwards and downwards. The wire is connected to an electrical pack with voltage adjustments (0–100 V). Upon supplying current to the hotwire, it is heated and can slice the large foam block into sheets of desired thicknesses.

- *Process*: the main steam lines are, in general, insulated and drawn through the accumulator to the moulds positioned at different stations. The branching steam supply lines to the moulds are not insulated to ensure saturated hot steam does not reach the moulds. The accent is on steam volumes and not on pressure. The moulds are filled using 'fill-guns' operated by air with excess material going back into the supply point. The steam activates the water-blowing agent inside each bead and expands to form one mass and takes the shape of the mould. This takes only a short time, after which the moulds are cooled and moulded product removed. This cycle is repeated. Insufficient cooling results in shrinkage, whereas premature opening of a mould results in a distorted form due to continuing expansion. If the blocks are vacuum-cooled, they can be cut into sheets in a short time or, if water-cooled, they need ≥24 h to mature (removal of moisture). The scrap generated by fabrication of the blocks into sheets can be shredded/granulated and ≈20% can go back each time a block is made.

8.6 Expandable polystyrene fillers in water-blown rigid polyurethane foams

Fillers are added to polymers to reduce costs, improve processing behaviour, or to modify the properties of the final product. Fillers are inexpensive but must be compatible with the other components being used, especially the grade of polyol or a polyol system. Traditionally, fillers were used to reduce costs, making the final product cheaper, but this is no longer the only reason (nor even the most important). Fillers influence density and other material characteristics such as: optical; surface; electrical, magnetic, mechanical and rheological properties; chemical reactivity; thermal stability; flame retardancy.

Fillers are also added to foam formulations, though their production parameters may be sensitive to processing. The inclusion of fillers can complicate a foaming process, and it is advisable when using non-traditional fillers to carry out in-house laboratory testing using a cup test or box test. Perhaps, instead of using one filler, a combination of fillers or a system may improve the quality of the processed foam. One advantage of using fillers is they reduce shrinkage and combustibility. Some of the common fillers used in the production of water-blown rigid PUF are aluminium hydroxide, calcium carbonate, melamine, starch, borax, and crystallised silica. I would like to add expanded graphite and biomass ashes to this list, especially wheat husks ash and rice hulls ash, both of which are rich in silica.

An interesting application is the production of PU/polystyrene porous composites. These porous composite materials (based on water-blown rigid PUF as the matrix and containing thermoplastic EPS beads as fillers) have some better properties thermal conductivity, compressive strength, core apparent density, and dimensional stability. The process involves expansion of the EPS beads within the PU mix, where the reaction between the polyol and isocyanate and the water produces gas, which is an exothermic reaction. The proportion of the EPS as a filler agent can be 20–40% of the total mix. Rigid PUF filled with EPS show good mechanical properties and dimensional stability. These foams are equal to (and sometimes better than) those without EPS. Such composites could be cheaper and have important industrial applications.

Bibliography

1. *SEALECTION® PIP*, Technical Data Sheet, Demilec Inc., Nuneaton, UK, 16th February 2015.
2. *Expanded Polystyrene Fillers in Water-Blown Rigid Polyurethanes*, Cracow University of Technology, Poland, 2011.
3. *Polyurethane Spray Foam Insulation: Polarfoam Soya*, Coast Spray Foam Ltd., Abbotsford, Canada.
4. *Self-Skinning Water-Blown Flexible Foams-S-Foam 48*, Technical Data Sheet, Barnes, Bankstown, NSW, Australia, 2010. http://www.kirkside.com.au/Uploads/Images/barnes-s-foam-48.pdf.
5. *Spray Foam Insulation: Water Blown Two-Component Systems*, JHMorrow, Elk Grove, CA, USA.

9 Recommendations for process efficiency

9.1 What is process efficiency?

Process efficiency is the capability of human resources to carry out a certain process in a way that ensures minimising consumption of effort and energy, materials and time to produce a quality product. A broad-based formula to measure efficiency is:

$$\text{Efficiency} = \text{actual output/standard output} \times 100\%$$

Where, actual output is the total amount of production achieved and standard output is the average historical performance of the same process.

In simple terms, as it appears from the formula, to increase the efficiency of a process, an organisation needs to focus on increasing the actual output. To do this, a full-scale analysis of the whole process is needed to find out the problem areas and then implement solutions to enhance the process.

Now that we have a good theoretical knowledge about water-blown cellular polymers and their processes, the next step is to discuss how to make these foams efficiently, safely and, of course, profitably. This chapter deals with this aspect of cellular foams.

In industrial ventures, it is important to plan effectively from concept to implementation and beyond to ensure a smooth and successful operation. This is particularly true if dealing with chemicals or manufacturing processes in which chemicals are involved. If the chemicals used are toxic or hazardous, extra precautions must be implemented.

In the processing of polymers and other chemicals to produce cellular polymers, certain basic standards must be met. To meet these standards, a manufacturer must put in place certain basic essentials: a well-planned factory; well-trained and knowledgeable supervisory staff; safety equipment and procedures; appropriate storage facilities for raw materials; emergency stations for first aid; a good ventilation system; an effective production process system on the floor; key factors on a production floor; in-house laboratory; quality-control system. Most of these features have been discussed in the preceding chapters, but the last three items are presented here.

9.2 Key factors on a production floor

Foam production generates exciting but rewarding challenges, but several factors influence a smooth, safe and profitable operation. To overcome problems, the following are recommended:

- *Mission statement*: company objectives, company vision, customer policy.
- *Well-planned factory layout*: good storage, good ventilation, good production flow.

https://doi.org/10.1515/9783110643121-009

- *Posters*: motivation, safety, warnings, hazards, correct procedures.
- *Machinery*: appropriate installation, operator training, maintenance, spares.
- *Process engineering*: time-and-motion study, constant process improvement.
- *Product development*: quality improvement, new products, cost cutting.
- *Quality assurance*: quality control, documentation, reporting, analysis, action.
- *Waste control*: minimise waste, waste recycling.
- *Production methods*: operating systems, correct procedure, correct tools.
- *Workforce*: good morale, enhance technical and operating skills, teamwork.
- *Raw materials*: material quality, good storage, correct formulations, safety.
- *Ventilation*: adequate ventilation, efficient exhaust system, air-quality checks.
- *Good management*: communication, motivation, incentives, good facilities.
- *Troubleshooting*: solving problems, spill management, excess waste, returns.
- *Fire precautions*: fire extinguishers, training, fire drill, emergency exits.
- *Tools and equipment*: appropriate training, inventory, issue systems, safety.
- *Employee facilities*: lunch room, washrooms, lockers, uniforms, parking.
- *Production floor committee*: manager, supervisors, lead hands, regular meetings.
- *Downtime*: minimise, eliminate causes, perhaps a standby generator.
- *Job rotation*: between hard/easy jobs for operators.
- *Absenteeism*: minimise, eliminate causes, attendance bonus.
- *Rivalries*: avoid rivalries between supervisors/operators/others.

In general, foam production generates wastes of ≈10–15%, which is acceptable. Most of these wastes are due to the 'skins' formed by the foam blocks. These can be shredded and compressed into foam blocks again using a steaming process or adhesives. These blocks can be cut into thick slabs for mattress bases or can be cut thinner and supplied for carpet underlay.

A foam operation can be monitored for performance periodically, comparing use of raw materials in kilogrammes *versus* the foamed output or raw materials used in kilogrammes *versus* the monetary value. These exercises will also show waste percentages and other data to enable a foam producer to take corrective action where necessary without waiting for the accounting statements (at which point it will be too late to take action). Another financial indicator that can be used to monitor foam production mid-stream is the productivity factor: the ratio of the value of the output divided by the value of the input. If it is <1, the operation is running at a loss.

9.3 In-house laboratory

Initially, an in-house laboratory does not have to be sophisticated. The main purpose of a mini-laboratory is to test different foam formulations in small quantities before they go on the production machines. The two main basic tests required are density and indentation force deflection (IFD) (support factor) and they can be done using two basic methods. The first method is a cup test. A small quantity of a formulated foam mix is poured into a medium-size plastic cup and allowed to overflow slightly. The foam is removed and the top excess cut-off, and then the small block of foam tested. The other, more accurate method is to pour a calculated amount of a foam mix into a wooden/other type of box of size 30 × 30 × 15 cm. From both tests, one can calculate the mixing time, pouring time, creaming time, full rise time, curing time (not tacky) and demoulding time. In this way, one can formulate differently to obtain different densities, IFD, and properties. These foam blocks should be cured for 24 h before carrying out density, IFD and other tests.

These pre-tests save a foam producer a lot of waste because he/she can modify formulations before they go on the production machines. This laboratory could start as a mini-laboratory, but a knowledgeable person must be in charge. The same person/persons can also carry out quality-control work.

9.4 Quality-control systems

Quality-control systems are essential for any business. Apart from running a business profitably, all customers expect quality assurance from a foam manufacturer. For bedding and furniture industries, customers look for at least conformity to their requirements of density and support factor (IFD). Other customers who order foams with special properties may request certification, and a foam producer should produce quality foams at all times.

There are different systems and standards to ensure quality. Probably the most popular system is statistical process control (SPC). Most foam manufacturers have their operations certified to International Organization for Standardization (ISO) standards such as ISO 9001, ISO 9002, and QS 9000 (automobile industry). Some of the popular standards are the British Standard (BS) and the American Society for Testing and Materials (ASTM). Once an organisation has been certified to these standards, customers accept quality without problems.

9.5 Lean processing for increased productivity

'Lean processing' or 'lean production' is a production practice that considers the application of resources for any operation, other than the creation of value for a product or service, to be wasteful and thus a target for elimination. Elimination of waste of resources in any form during each process cycle results in enhancement of productivity and, thus, collectively results in increased profitability. From the perspective of a customer who 'consumes' a product or service and is paying for this 'value', it can be defined as any action or process for which a customer is willing to pay for. One might say that lean processing is centred around increased value at less cost with quality in mind at all times.

Lean practices can be applied to any sector of an operation of a business or industry and, if applied effectively, can elicit great surprises and possibilities. The methods of waste reduction are challenging, exciting and rewarding. Waste does not mean only material waste but applies to all areas of an operation, such as from management to finished products to the final point of shipping. General lean methodology is based on three types of waste reductions: no-value-added work, overburdening a process, and uneven process flow.

A lean operation is a variation on the theme of efficiency and is based on optimising smooth flow. If applied to a manufacturing industry, a lean operation can be considered to be improving the efficiency of a process and decreasing waste in all aspects of the process. It could be called 'fine-tuning' of a process. Lean practices provide a set of 'tools' that assist in the identification and steady elimination of waste to achieve pre-set goals. As waste is eliminated, the quality of value improves, while production time and costs are reduced. There is a second method for lean manufacturing in which the main focus is improving the overall 'flow' or smoothness of a process, thereby eliminating the 'unevenness' of work, operation or a system.

9.6 Industrial engineering

Industrial engineering is a branch of engineering that deals with optimisation of complex processes or systems. Industrial engineers work to eliminate waste of time, materials, man-hours, machine time, energy and other resources that do not generate value, and this invariably equates to money. These engineering processes and systems improve quality and productivity.

Industrial engineers are professionals trained to increase productivity specifically through study and improvement of processes. Some of the tools they use are time management, SPC, quality-assurance systems, operator behaviour studies, process flow data, standards and target vectors, and work study. Industrial engineering deals with technical and business aspects.

Industrial engineering is concerned with the study, development, improvement and implementation of integrated systems of people, money, knowledge, information, equipment, energy, materials, analysis and synthesis, as well as the mathematical, physical, chemical factors together with the principles and methods of engineering design to evaluate, specify, predict and establish methods or systems to improve processes. Depending on the sub-specialities involved, industrial engineering may also be associated with operations management, management science, operational research, systems engineering, manufacturing engineering, human-behaviour patterns, safety engineering and utilise any of these combinations to meet a particular need for a given process.

Perhaps 'industrial engineering' should be called 'manufacturing engineering' or 'production engineering' or, better still, 'systems engineering' because it encompasses any methodical or quantitative approach to optimising how a system, process or organisation operates.

The topics involved in industrial engineering are:

- *Accounting*: analysis of data, measurement, processing of financial information.
- *Management science*: disciplines of information, methods to make effective decisions.
- *Operations management*: designing, overseeing and control of processes.
- *Job designs*: job contents, variations for improvement, effective scheduling.
- *Project management*: planning, organising, motivating, procedures, targets.
- *Engineering management*: applying engineering principles to business practice.
- *Supply management*: storage of goods, flow of goods, work-in-progress, finished goods to consumer.
- *Process engineering*: operation, control, optimisation of chemical, physical and biological processes.
- *Systems engineering*: focuses on machinery and systems designs to maximise output efficiencies and their lifecycles.
- *Ergonomics*: practice of designing products, appropriate accounting of systems or processes as well as the interaction between them and the people who use them.
- *Safety engineering*: engineering disciplines to ensure safety to acceptable standards.
- *Cost engineering*: management of project costs, analysis of utilities, risk analysis.
- *Value engineering*: systematic method to improve the 'value' of products or goods.
- *Quality engineering*: quality assurance, process quality control, recording, analysis and action to assure quality in all areas of production.
- *Facility management*: coordination of space, infrastructure, people and organisation.
- *Logistics management*: flow of goods from delivery, storage, issue, usage, wastage, disposal, order levels, and transport.

Previously, a major aspect of industrial engineering was planning factory designs and layouts, designing assembly lines, maintenance systems and other manufacturing requirements. Today, especially if lean manufacturing systems are used to enhance process control, quality assurance, waste reduction, and plant efficiency to increase productivity and profitability, industrial engineers work to eliminate waste of time, money, materials, energy and other resources.

Examples of where industrial engineering might be used include: charting of flow processes; process mapping; designing an assembly workstation; strategising for various operational logistics, quality systems, recycling systems, transport and shipping. Industrial engineers use systems such as time-and-motion studies, work studies, computer simulation, as well as extensive mathematical tools for proto-types, computer analysis, evaluation and process optimisation.

9.7 Maintenance of work equipment

For any type of process efficiency, all related work equipment must be in prime working order. This can be achieved by a combination of preventive maintenance and routine maintenance. Before thinking of maintenance, one must fully understand the operating systems, required safety measures, and limitations of machinery equipment. The machinery suppliers provide full data and, in all probability, also carry out the installations. Before they leave, the engineering staff of the foam producer must be thoroughly familiar with their operations, shortcomings (if any), lifespans and type of maintenance required.

Factory owners must ensure that work equipment does not deteriorate to the extent that they may put operators at risk, Accidents may happen but they should not be due to faulty machines. The frequency and nature of maintenance should be determined based on efficiency and risk assessment taking full account of:
- Recommendations of the equipment manufacturer
- Intensity of use
- Operating environment
- Knowledge and experience of the user
- Risks to health and safety from failure or malfunctions
- Risks to health of working with hazardous chemicals

As a rule, all moving parts of equipment must be provided with guards, protection from electrical shocks, and all operators must be trained fully in operation of equipment. Some of the critical parts may need a higher and more frequent level of attention than other aspects, which should be highlighted within the maintenance programme that has been implemented. It is advisable to have a maintenance recording log at each machine and workstation. Having these alone is not sufficient but the engineering staff must carry out regular checks of the operators comments and action taken must also be recorded. This should be in addition to the recordings by staff of routine preventive maintenance carried out.

Preventive maintenance can be based on experience of using such machines and also guided by the recommendations of the equipment supplier. Perhaps you may work out a better system than that has been recommended. Listen to what the operators have to say and also check what they have logged because they are the best source of knowledge. In general, serious maintenance work is carried out during factory shutdowns (e.g., holiday period or long weekend). If machinery and equipment are new, maintenance becomes easy but, as their lifespan (in general, 10 years) comes closer, more attention is needed.

9.8 Foam production

Foam productions may take the form of small-volume, medium-volume or very-large-volume production. The methods employed are manual, semi-automatic (block by block) or continuous foaming (in which foam production is a continuous large block of foam on a slowly moving conveyor and cut to size). Irrespective of the method used, cooler room temperatures ($\approx$20 °C) give rise to the best qualities. Some foam producers may opt to have the foaming area air-conditioned if foaming plants are located in hot/warm climates. The following recommendations aim to improve foam yields and minimise waste:

- *Manual operation* is where small foam blocks are made using an electric drill or a hand-held mixer. Production volumes are small but good-quality foam can be made if formulated appropriately and weighing is accurate. Two important aspects before production are a box test and a floating lid to flatten the rising foam before it forms a meniscus. The mould design should minimise the thickness of the 'skins' that form on all six sides of a block.
- *Semi-automatic or intermittent process* is where large-size foam blocks can be made one at a time. Many foaming machinery systems are available and, in this case, components can be metered accurately into a central mixing head to preset quantities. A box test and a floating lid help to minimise waste. Detachable moulds should be designed so as to make the skins as thin as possible to minimise waste. Allowances should be made for material loss due to escaping gas, shrinkage, and losses due to skins.
- *Continuous foaming*: these systems can produce very large volumes. The continuous flow of foam cures itself on the conveyor by the time it reaches the online cutting systems, which cut them into 'buns/blocks' of desired sizes. Usually, these blocks are cut into arbitrary or convenient sizes and, when fabricated into final products, result in unusable foam waste. A foam producer knows the exact sizes of the products (e.g., mattresses, cushions) so these foam blocks can be cut to the exact, no-waste sizes.

Most continuous foaming lines do not have a 'floating-lid system' but modern lines do, and 'flat-top' blocks can be made to save a lot of foam waste.

9.8.1 Case study 1: Hands-on solution by the author

A large foam manufacturer was experiencing excessive waste (35%). A continuous foaming line did not have a 'floating device' and the foam blocks had rounded tops. The foam blocks were also cut into convenient sizes but equal in length. Material wastage at the start and end was normal, in accordance with usual practice.

Operators were highly skilled and cooperative. The crew came in at 7 am for work but the foaming was done after 10.30 am. The foaming run was short. The target was to reduce wastage to 15%

After observation and study of processes for 3 days, the following was implemented:

- The crew to clean the machine and set it up before they left, ready for production the next day. No need to waste time by cleaning/setting up in the morning.
- Production started at 7.30 am when it was cooler and not after 10.30 am (warmer).
- Production run was extended by three times of the volume: longer run, same waste.
- Foam blocks were cut to sizes as advised by the production department on the previous day.
- Semi-cooled foam blocks were taken to the curing room and set-up upside down to flatten tops.
- Each block was marked as per the instructions from the production department for easy identification.
- Foam wastes from 'ends' (start/finish) were large, so production was able to cut cushions before sending the balance wastes for the shredding section.
- Storage of foam blocks based on first-in/first-out system. Ensure full cure.
- With these basic changes, the foam producer was able to reduce waste to 12%, which is acceptable.

9.8.2 Case study 2: Hands-on solution by the author

Foam producer making thin foam sheeting in roll form for mattress padding, the objective to reduce waste and make foam cells finer for easy 'peeling'.

Observation of production processes and peeling operation revealed that foam cells were coarse and the starting block was in the form of a rectangular block. The formulation was examined and modified with increase in water content (blowing agent) and air intake was increased to make foam lighter and to produce fine cells. The foam producer was advised to make the blocks for this process in 'round' blocks instead of rectangular blocks, which produced unnecessary waste. The finer 'texture' made the peeling operation much easier and allowed the making of thinner sheeting than before.

Process study and constant improvements in a foam operation are exciting challenges. With newer products (especially additives) and fillers constantly appearing in the market, improvements are that much easier.

9.9 Basic indicators to monitor efficiency

Businesses aim to operate at a profit. Foam productions are big-volume operations with a well-established global market. With population increases, the demand for foam products will also increase. Foam manufacturers, especially large-volume producers, have invested large amounts of money and want to ensure good profit margins, so their processes must run efficiently. Given below are a few basic indicators that can help management to plan corrective action where necessary.

9.9.1 Breakeven point

Most foam productions result in small or large foam blocks. Even if a continuous process is used, continuous foam is cut into 'buns' and ends up as foam 'blocks'.

For example, a foam manufacturer makes foam blocks. Each block will sell for US$100. It costs US$40 to make each one and the fixed cost for that period is US$15,000. How many blocks must be made to break even and start making profit?

Apply formula:

$$X = FC/(SP - VC) = 15,000/(100 - 40) = 250$$

Where:
  X: number of blocks;
  FC: fixed costs;
  SP: selling price; and
  VC: variable costs.
  Therefore, the foam producer must make 250 blocks to break even.

9.9.2 Return per kilogramme

The return per kilogramme is the total gross value of sales (or sale value) of goods *versus* the total raw material consumed in kilogrammes during that period. For example, a foam manufacturer makes 1,000 mattresses and each sells for US$115. The total raw materials consumed, including waste, is 1.200 kg, then:

$$Return/kg = (115 \times 1,000 \div 1200) = US\$95.83$$

Compare with pre-set standard.

9.9.3 Cost per kilogramme

Total gross cost of operation *versus* total cost of raw material consumed during a given period:

Total costs = Ex-factory + marketing + administration + other = US$72,000

Total raw material consumed during this period = 1,200 kg
Therefore:

$$\text{Cost per kg} = (\text{US\$72,000} \div 1,200 \text{ kg}) = \text{US\$60.00}$$

Compare with pre-set standards.

Bibliography

1. *The Safe Maintenance Health Check*, Health and Safety Executive, Bootle, Merseyside, UK. http://www.hse.gov.uk/safemaintenance/checklist.htm.
2. *European Campaign on Safe Maintenance*, European Agency for Safety and Health at Work, Bilbao, Spain.http://osha.europa.eu.
3. C. Defonseka in *Practical Guidelines to Flexible Polyurethane Foams*, Smithers Rapra, Shawbury, Shropshire, UK, 2013.

10 Recycling of cellular foam wastes

In the world of polymer cellular foam, ≈80% comprises polyurethane foams (PUF) with closed or open cells. Thus, this chapter focuses mainly on PUF. Processing polymers into plastics products generates wastes irrespective of the methods used. However, we are familiar with the disposal or recycling methods of most plastics, and PUF can also be recycled effectively.

Like other plastics, many polyurethane (PU) products can be recycled in various ways to remove them from the waste stream and to recapture the value inherent in the material. Most consumers are familiar with recycling of plastic bottles and containers and there are post-consumer programmes in place from collection to transport to central recycling centers. In fact, this is a big business because millions of tons are available for recycling, which ends up as pellets routed back for production. PU recycling usually happens on production floors and in industrial settings. It takes many forms, from relatively simple shredding to breaking it into its chemical constituents.

10.1 Polyurethane recycling

A few examples of different types of PU recycling are:
- *Rebond*: according to statistical analyses, nearly £1 billion of reclaimed PU scrap was used in 2010 to create rebond cushioning used as carpet underlay.
- *Mattresses*: large-volume foam producers can generate ≤350 kg/day, which is recycled.
- *Raw material from recycled material*: a manufacturer in the USA provides raw materials for PU production (polyols) with ≈70% recycled content that are used in the automotive industry.

10.1.1 Polyurethane recycling processes

PU are recycled in two primary ways: i) mechanical recycling (the material is reused in its polymer form); and ii) chemical recycling (takes the material back to its various chemical constituents).
- *Rebonded flexible foam*: rebounded flexible foam or 'rebond' is made if foam scrap is shredded and an adhesive (binder) is used to create carpet underlay, sport mats, cushioning and similar products. Another process is to soften the shredded foam scrap by steam and press it into well-bonded large blocks from which sheets and slabs can be cut for many applications (including carpet underlay). Rebonded material has been used for decades and represents ≈90% of foam used as carpet underlay.

https://doi.org/10.1515/9783110643121-010

- *Regrind or powdering*: post-industrial PUF scraps are ground into a fine powder. The resultant powder is mixed with virgin material or can be used as a filler in foam formulations. Ideal for reaction injection moulded (RIM) parts.
- *Adhesive pressings/particle bonding*: these two processes use PUF scrap from various applications (e.g., bedding, automobile parts, refrigerator and industrial trim) to create boards, and mouldings, often with very high recycled content. Used PU parts are granulated and blended with a powerful binder or PU systems and then formed into boards or mouldings under heat and pressure. The resulting board is similar to particle board made from wood waste, and is used in sound-proofing applications and furniture that are virtually impervious to water and flooring where elasticity is needed.
- *Compression moulding*: this recycling process grinds RIM and reinforced parts into fine particles and then applies high heat and pressure and heat in a mould, creating products with ≤100% recycled content and material properties that can be superior to those of virgin materials.

10.1.2 Chemical recycling

A few examples of different types of chemical recycling are:
- *Glycolysis* combines mixed industrial and post-consumer PU with diols at high heat, causing a chemical reaction that creates new polyols the basic raw material for making PU. These polyols can retain the properties and functionality of the original polyols and can be used in several applications.
- *Hydrolysis* creates a reaction between used PU and water, resulting in polyols and various intermediate chemicals. The polyols can be used as fuel and the intermediates as raw materials for PU.
- *Pyrolysis* is where the reaction breaks down PU under an oxygen-free environment to create gas and oils.
- *Hydrogenation* is similar to pyrolysis. Hydrogenation creates gas and oil from used PU through a combination of heat, pressure and hydrogen.

10.2 Key opportunities to expand recovery of polyurethane foam

Some of the main uses of recycled PUF are carpet underlay, rebounded slabs as mattress bases, padding in furniture, thin sheets for packaging, and padding for cycle seats. Also, a percentage of the shredded foam waste on a production floor can go back into production. The market for 'virgin foam' flooring for carpet underlay is small due to the improved value and increased sustainability of rebonded foam. In addition to these markets, rebonds are making inroads in hotels, institutional, retail and even marine applications. Some use it for acoustic insulation in automobiles, gym pads, prayer mats, and pet bedding. The cheaper cost factors also help influence these applications.

Consumers have a considerable impact on the success of recycling of any material. By actively collecting products for recycling and by buying products that are recycled or made with recycled content, consumers help drive supply and demand. In some instances, consumers can make a direct choice to buy products made with recycled PU. For example, if buying carpeting, consumers can specify carpeting with recycled PU. To encourage recycling and manufacture of quality products, consumers can seek out retailers or manufacturers that use recycled wastes.

10.3 Some value-added products from foam wastes

Foam wastes can also be converted to value-added products for important applications such as adhesives, coatings, and varnishes. There are many possibilities, so the section below is confined to three of the important areas of applications.

10.3.1 Adhesives from foam wastes

Reuse of polymer waste is a goal of many sustainability initiatives. Much work has been done but most of it has been commodity plastics such as consumer products and packaging. Scrap polymeric materials, including those left over from consumption, as well as those left over from production but which are not useful for various reasons, comprise a vast range of reclaimable material for potential conversion to other value-added products. In the adhesive area, activity has focused mainly on reuse of wastes from PU, polyester and cellulose.

Adhesives consume non-renewable resources such as petroleum and they contribute to environmental pollution and the waste stream. The latter can include residues of cured or uncured adhesives or other polymers. Waste is a well-known contributor to production costs with attached health and safety issues. In all approaches to sustainability, there is a sequence of choices in decreasing order of desirability. Reusing materials that otherwise might be introduced into the waste stream is the next most environmentally and economically beneficial strategy. Disposal in any form, particularly in landfills, is the least desirable strategy in that it takes up space, is costly, and could result in harmful emissions and contaminants. The possible use of foam wastes for making adhesives needs to be focused upon, in several ways:

- Reuse of relatively pure foam scrap
- Reuse of polymer mixtures
- Burning of foam scrap to recover its energy content
- Chemical conversion of foam wastes to recover useful materials
- Separation of polymers from mixed-foam waste streams

PUF wastes can also be processed *via* glycolysis and used as raw material in the production of a two-component, general-purpose adhesive. The two-phase liquid resulting from glycolysis is the polyol component in the adhesive and is reacted with diphenyl methane diisocyanate. The glycolysis mixture can be used without side products and thus, complete recycling of the PU waste can be achieved. This adhesive provides good bonding strength to wood, metals, plastics, glass, ceramics and leather. It has been found that foam wastes from polyethylene glycol adipate urethanes can be plasticised with dimethylformamide to produce an adhesive compound with properties similar to those of epoxy adhesives. Foam wastes can be formulated in many ways depending on end-applications basic to industrial needs to more specialised applications in which high bonding strengths and weathering properties are important.

10.3.2 Coating applications

Recycled polyols are derived from PU soft foams by a combination of glycolysis and aminolysis. In this process, cleavage of urethane bonds results in urea groups and hydroxyl compounds. The polyureas from the water reaction of polyisocyanate are, in general, not cleaved due to short reaction times and moderate temperatures. By this process, homogenous polyols of low glycol content are obtained, the hydroxyl number of which is adjusted to achieve the desired properties of the coatings, which generally are in the range 180–300 mgKOH/g. The viscosity of the recycled polyol would probably be in the range 3,000–6,000 mPa.s at 25 °C with some influence of the amount of dissolved polyureas originating from the reaction mentioned above.

Coatings can be produced by simple mixing of the recycled polyol plus additives in low concentrations and with diisocyanate or polyisocyanate. Hardness and elasticity of the coatings can be adjusted by varying the isocyanate index. The coatings prepared thus have better tensile strength and better elongation properties than most trade products. These coatings can be sprayed by spray guns onto concrete, paper, sheet metal, plastic films or can be applied by knife-coating if very thin coatings are desired. They are transparent or with a slightly yellowish tint, and can be coloured using dyes.

10.3.3 Polyurethane varnishes

In general, varnishes are transparent, hard, protective finishes used primarily for wood and other materials. A varnish is a combination of a drying oil, a resin, and a thinner/solvent. Varnish finishes are usually glossy but may be designed to achieve satin, semi-gloss or matt finishes by the addition of 'flatting' agents. They have little or no colour, are transparent and pigments/dyes can be added to produce any colour desired. For very-high-gloss finishes, appropriate additives can be added. Varnishes are also applied over wood for staining for aesthetic effects and as a film for gloss and protection. Some products are marketed as a combined stain and varnish.

PU varnishes are typically hard, abrasion-resistant and durable coatings. They are popular for hardwood floors but are considered by some furniture manufacturers to be difficult for finishing furniture or other detailed pieces. PU are comparable in hardness with certain alkyds but generally can form a tough film. Compared with simple oil- or shellac-based varnishes, PU varnishes form harder, tougher and more waterproof films. However, a thick film of PU tends to delaminate if subjected to heat or shock for long periods of time, leaving white patches. This tendency increases with long periods of exposure to sunlight or if they are applied over soft woods (e.g., pine). This effect is due mainly to PU varnishes having less penetration into the applied surfaces. This can be overcome by additives or using various priming methods, including use of certain oil varnishes, specified 'dewaxed' shellac, clear penetrating epoxy sealer or 'oil-modified' PU designed for this purpose. However, PU varnishes have the advantage of a much faster build-up of thicker coats than possible with other varnishes to reach the desired finished effects.

Unlike drying oils and alkyds which cure after evaporation of the solvent upon reaction with oxygen in the air, PU coatings cure by various reactions among the chemicals in the mixture or instant curing with moisture in the air. Certain PU products are 'hybrids' and combine different aspects of their parent components. 'Oil-modified' PU (water- or solvent-borne) are used widely for wood floor finishes. Use of PU varnishes can cause problems for exterior applications due to their susceptibility to deterioration if exposed to ultraviolet (UV) light, and the incorporation of UV absorbers and pigments help to a great extent. PU are very versatile materials (especially the flexible foam wastes), so they can be converted to many value-added products, including lacquers and wood preservative coatings.

Appendix 1   Suppliers of raw materials and foaming machines

Some sources for raw materials and foaming machines are shown below. The reader may also research to find other sources to suit particular processes.

A1.1 Raw materials

- Dow Corporation (USA)
- BASF (Germany)
- Bayer AG (Germany)
- Huntsman Corporation (USA)
- Chemcontrol Limited (USA)
- Issac Industries Incorporated (USA)
- ERA Polymers Limited (Australia)
- Union Carbide Limited (Canada)
- Premilec Incorporated (Canada)
- Bio Based Technologies LLC (USA)

A1.2 Foaming machinery

- Hennecke GmbH (Germany)
- Canon Viking (UK)
- Beamech Limited (UK)
- Laader Berg AS (Norway)
- AS Enterprises (India)
- Modern Enterprises (India)
- Edge Sweets Company (USA)
- Elitecore Machinery Limited (China)
- Sunkist Machinery Company (Taiwan)
- Dongguan Hengsheng Company (China)
- Mane Electricals Limited (India)

There are many other suppliers for raw materials and foaming machinery from other countries such as Japan, Italy, and the Netherlands. For working with water-blown expandable polystyrene for which a specific pre-expander is required, Mane Electricals Limited of India specialise in expandable polystyrene machinery and can be recommended. If selecting suitable foaming machines (especially for polyurethanes) one may opt for two-component systems depending on the volume of production.

https://doi.org/10.1515/9783110643121-011

Appendix 2 Conversion table of units commonly used in industry

	To convert	Into	Multiply by
Airflow	ft^3/min	litre/s	0.4719
	ft^3/min	m^3/s	0.0004719
Coating	g/m^2	oz/yd^2	0.0295
	ounce/yd^2	ounce/m^2	33.90
Density	lb/ft^3 (pcf)	kg/m^3	16.018
	kg/m^3	pcf	0.0624
	pcf	g/cm^3	0.016
	g/cm^3	pcf	62.43
	g/L	pcf	0.0624
Energy	joule	ft-lb	0.7573
	joule	in-lb	8.85
	ft-lb	joule	1.355
	in-lb	joule	0.113
	Btu	joule	1.055×10^3
Indentation load deflection or indentation force deflection (load-bearing)	lb/50 in^2	N/323 cm^2	4.448
	N/323 cm^2	lb/50 in^2	0.225
	lb/50 in^2	kg/323 cm^2	0.4536
	kg/323 cm^2	lb/50 in^2	2.2
Length	angstrom	metre	1×10^{-10}
	metre	micron	1×10^6
	micron	angstrom	1×10^4
Pressure and stress	lb/in^2 (psi)	(kpa)kN/m^2	6.895
	kN/m^2 (or kPa)	lb/in^2	0.145
	kg/cm^2	kPa	98.07
	kg/cm^2	psi	14.223
	psi	kg/cm^2	0.0703
	psi	Pa	6895
	Pa	psi	0.000145
	g/cm^2	psi	0.01422
	psi	dyne/cm^2	68965

(continued)

(continued)

	To convert	Into	Multiply by
	dyne/cm^2	psi	0.0000145
	bar	atm	0.987
	bar	kg/cm^2	1.02
	bar	Pa	1×10^5
	dyne/cm^2	atm	9.869×10^{-7}
	psi	atm	0.068
	Pa	dyne/cm^2	10
	MPa	psi	145
	atm	Pa	1.013×10^5
	dyne/cm^2	Pa	0.1
Tear strength	lb/in	N/cm	1.75
	N/cm	lb/in	0.571
	N/m	lb/in	0.00571
	lb/in	N/m	175.1
	lb/in	kg/cm	0.1786
	kg/cm	lb/in	5.6
	kN/m	N/cm	10
	kg/cm	N/cm	9.798
Temperature	°C	°F	9/5 (°C) + 32
	°F	°C	(5/9) (°F − 32)
	°C	K	°C + 273.15
Volume	litre	in^3	61.023
	litre	US gallon	0.264
	litre	ft^3	0.0353
	US fl oz	ml	29.6
	US fl oz	m^3	2.957×10^{-5}
	US gallon	litre	3.79
	litre	m^3	0.001
	m^3	litre	1,000
	litre	US fl oz	33.819
	US fl oz	British fl oz	1.0408
	Board foot	m^3	2.359×10^{-3}

(continued)

(continued)

	To convert	Into	Multiply by
Viscosity	Centipoise = centistrokes × density		
	Pascal second (Pa.s)	centipoise	1,000
	centipoise	Pa.s	0.001
	mPa.s	centipoise	1
Weight	g	metric tonnes	1×10^{-6}
	g	kg	0.001
	g	ounces (avdp)	0.0352739
	g	lb (avdp)	0.0022026
	ounces (avdp)	g	28.3495
	ounces (avdp)	metric tonnes	2.83495×10^{-5}
	metric tonnes	kg	1,000
	metric tonnes	lb	2,240

IFD: Indentation force deflection

Abbreviations

2D	Two-dimensional
3D	Three-dimensional
ABS	Acrylonitrile–butadiene–styrene
ASTM	American Society for Testing and Materials
BS	British Standards
CBA	Chemical blowing agents
CFC	Chlorofluorocarbon(s)
CO_2	Carbon dioxide
EPS	Expandable polystyrene
EW	Equivalent weight
HDPE	High-density polyethylene
IFD	Indentation force deflection
ISF	Integral skin foams
ISO	International Organization for Standardization
JIS	Japanese Industrial Standards
LDPE	Low-density polyethylene
LLDPE	Linear low-density polyethylene
MDI	Diphenyl methane diisocyanate
MW	Molecular weight(s)
NIPF	Non-isocyanate polyurethane foam
NOP	Natural oil polyols
OH#	Hydroxyl number
OHV	Hydroxyl value
PA	Polyamide(s)
PBA	Physical blowing agents
PC	Polycarbonate
PCRH	Polymeric composites with rice hulls
PE	Polyethylene(s)
PET	Polyethylene terephthalate
PETE	Polyethylene terephthalic ester
PP	Polypropylene(s)
PPG	Polypropylene glycol
PS	Polystyrene(s)
PTFE	Polytetrafluoroethylene
PU	Polyurethane(s)
PUF	Polyurethane foam(s)
PVC	Polyvinyl chloride
QC	Quality control
SBO	Soybean oil
SPC	Statistical process control
SPF	Spray polyurethane foam(s)
SPP	Soy–phosphate polyol
TDI	Toluene diisocyanate
TEDA	Triethylenediamine
T_g	Glass transition temperature
TPE	Thermoplastic elastomer(s)

https://doi.org/10.1515/9783110643121-012

UF	Urea formaldehyde
uPVC	Unplasticised polyvinyl chloride
UV	Ultraviolet
WEPS	Water-blown expandable polystyrene
WPC	Wood–polymer composites
WPU	Waterborne polyurethane(s)
WVP	Water vapour permeability
XPS	Extruded polystyrene

Index

https://doi.org/10.1515/9783110643121-013

www.ingramcontent.com/pod-product-compliance
Lightning Source LLC
Chambersburg PA
CBHW081537220326
41598CB00036B/6467